PIC

PIC

Your Personal Introductory Course

John Morton

OXFORD BOSTON JOHANNESBURG MELBOURNE NEW DELHI SINGAPORE

Newnes
An imprint of Butterworth-Heinemann
Linacre House, Jordan Hill, Oxford OX2 8DP
225 Wildwood Avenue, Woburn, MA 01801-2041
A division of Reed Educational and Professional Publishing Ltd

ℛ A member of the Reed Elsevier plc group

First published 1998
Reprinted 1999

British Library Cataloguing in Publication Data
A catalogue record for this book is available from the British Library

Library of Congress Cataloguing in Publication Data
A catalogue record for this book is available from the Library of Congress

ISBN 0 7506 3932 6

Typeset by Avocet Typeset, Brill, Aylesbury, Bucks
Printed in Great Britain by Biddles Ltd, Guildford and King's Lynn

Contents

Acknowledgements

Max Horsey, Head of Electronics at Radley College in Abingdon and great driving force for technological advancement, first introduced me to PIC in 1995. With the help of Philip Clayton, now reading Computer Studies at Balliol College, Oxford, I was shown a new concept in circuit design – one which would radically change the way I saw the field of electronics.

I would like to take this opportunity to thank all those who have contributed, directly or indirectly, to make this book possible. First I must thank Richard Morgan, Warden of Radley College, for persuading me to try and get published, and my parents for their continual support with it. Chris Morton, my brother, was an invaluable proof-reader and I must also thank Pear Vardhanabhuti who started out with no knowledge of programming, and bravely took on the task of learning all about PIC using just the book. He then went on to design and build the 'diamond brooch' project circuit board. Also helping to build projects were Ed Brocklebank, James Bentley and Matt Fearn, and Matt Harrison helped me with the artwork involved. My work was greatly facilitated by Philip Clayton, an immaculate technical proof-reader and advisor. Finally comes the most important thanks of all, to Max Horsey – a constant provider of assistance and advice, and fountain of new ideas; he has helped me immeasurably.

1
Introduction

It has now become possible to program microchips; gone are the days when circuits are built around chips, now we can build chips around circuits. This technology knows no bounds and complex circuits can be made many times smaller. There is, however, little point in using PIC for a simple circuit that would, in fact, be cheaper and smaller *without* PIC, but most complicated logic circuits will benefit immensely by the use of PIC. By the way, PIC stands for peripheral interface controller.

When you buy a PIC, you get a useless lump of silicon with amazing potential. It will do nothing without – but almost anything with – the program that you write. Under your guidance, almost any number or combination of normal logic chips can be squeezed into one PIC program and thus in turn, into one PIC. Figure 1.1 shows the steps in developing a PIC program.

PIC programming is all to do with numbers, whether binary, decimal or hexadecimal (base 16; this will be explained later). The trick to programming lies in making the chip perform the designated task by the simple movement and processing of numbers.

What's more, there is a specific set of tasks you can perform on the numbers – these are known as instructions. The program uses simple, general instructions, and also more complicated ones which do more specific jobs. The chip will step through these instructions one by one, performing hundreds of thousands every second (this depends on the frequency of the oscillator it is connected to) and in this way perform its job. The numbers in the PIC can be:

1. **Received** from inputs (using an input 'port')
2. **Stored** in special compartments inside the chip (these are called 'file registers')
3. **Processed** (e.g. added, subtracted, ANDed, etc.)
4. **Sent out** through outputs (using an output 'port')

That is essentially all there is to PIC programming ('great' you may be thinking) but fortunately there are certain other useful functions that the PIC provides us with such as an on board timer (e.g. TMR0) or certain flags which indicate whether or not something particular has happened, which make life a lot easier.

The first section of this book will (hopefully) enable you to use the PIC16C54 and 55. These are two fairly simple PICs, and knowledge of how to

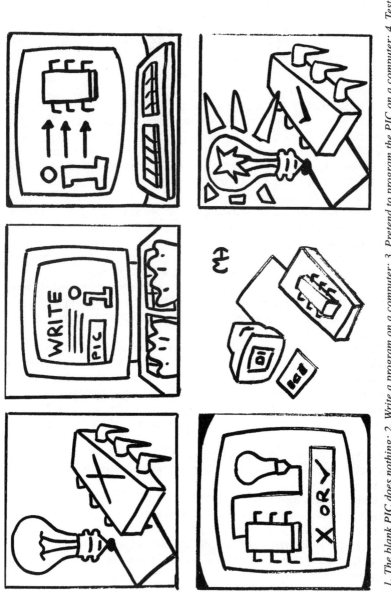

Figure 1.1 1. The blank PIC does nothing; 2. Write a program on a computer; 3. Pretend to program the PIC on a computer; 4. Test the program on a computer; 5. Program a real PIC; 6. Test the PIC in a real circuit.

use them will serve as a solid foundation to move on from, as there are many other diverse and exciting PICs around, and indeed new ones coming out all the time.

Some tips before starting

For those not familiar with programming at all, there may be some ideas which are quite new, and indeed some aspects of the PIC may seem strange. Some of the fundamental points are now explained.

Binary, decimal and hexadecimal

First there is the business of different numbering systems: binary, decimal and hexadecimal. A binary number is a *base 2* number (i.e. there are only two types of digit (0 and 1)) as opposed to decimal – *base 10* – with 10 different digits (0 to 9). Likewise hexadecimal represents *base 16* so it has 16 different digits (0, 1, 2, 3, 4, 5, 6, 7, 8, 9, A, B, C, D, E and F). Table 1.1 shows how to count using the different systems.

Table 1.1

Binary (8 digit)	Decimal (3 digit)	Hexadecimal (2 digit)
00000000	000	00
00000001	001	01
00000010	002	02
00000011	003	03
00000100	004	04
00000101	005	05
00000110	006	06
00000111	007	07
00001000	008	08
00001001	009	09
00001010	010	0A
00001011	011	0B
00001100	012	0C
00001101	013	0D
00001110	014	0E
00001111	015	0F
00010000	016	10
00010001	017	11
etc.		

The binary digit (or *bit*) furthest to the right is known as the least significant bit or *lsb* and also as *bit 0* (the reason the numbering starts from 0 and not from 1 will soon become clear). Bit 0 shows the number of 1s in the number. One equals 2^0. The bit to its left (*bit 1*) represents the number of 2s, the next one (*bit 2*) shows the number of 4s and so on. Notice how $2 = 2^1$ and $4 = 2^2$, so the bit number corresponds to the power of two which that bit represents, but note that the numbering goes from right to left (this is very often forgotten!). A sequence of 8 bits is known as a byte. The highest number bit in a binary word (e.g. bit 7 in the case of a byte) is known as the most significant bit (*msb*).

So to work out a decimal number in binary you could look for the largest power of two that is smaller than that number (e.g. 32 which equals 2^5 or 128 $= 2^7$), and work your way down.

Example 1.1 Work out the binary equivalent of the decimal number 75.

Largest power of two less than $75 = 64 = 2^6$. Bit 6 = 1	
This leaves $75 - 64 = 11$	32 is greater than 11 so bit 5 = 0,
	16 is greater than 11 so bit 4 = 0,
	8 is less than 11 so bit 3 = 1,
This leaves $11 - 8 = 3$.	4 is greater than 3 so bit 2 = 0,
	2 is less than 3 so bit 1 = 1,
This leaves $3 - 2 = 1$	1 equals 1 so bit 0 = 1.

So **1001011** is the binary equivalent.

There is however an alternative (and more subtle) method which you may find easier. Take the decimal number you want to convert and divide it by two. If there is a remainder of one (i.e. it was an odd number), write down a one. Then divide the result and do the same writing the remainder to the *left* of the previous value, until you end up dividing one by two, leaving a one.

Example 1.2 Work out the binary equivalent of the decimal number 75.

Divide 75 by two.	Leaves 37, remainder **1**
Divide 37 by two.	Leaves 18, remainder **1**
Divide 18 by two.	Leaves 9, remainder **0**
Divide 9 by two.	Leaves 4, remainder **1**
Divide 4 by two.	Leaves 2, remainder **0**
Divide 2 by two.	Leaves 1, remainder **0**
Divide 1 by two.	Leaves 0, remainder **1**

So **1001011** is the binary equivalent.

Exercise 1.1 Find the binary equivalent of the decimal number 234.

Exercise 1.2 Find the binary equivalent of the decimal number 157.

Likewise, bit 0 of a hexadecimal is the number of ones ($16^0 = 1$) and bit 1 is the number of 16s ($16^1 = 16$) etc. To convert decimal to hexadecimal (it is often abbreviated to just "hex") look at how many 16s there are in the number, and how many ones.

Example 1.3 Convert the decimal number 59 into hexadecimal. There are three 16s in 59, leaving $59 - 48 = 11$. So bit 1 is 3. 11 is B in hexadecimal, so bit 0 is B. The number is therefore **3B**.

Exercise 1.3 Find the hexadecimal equivalent of 234.

Exercise 1.4 Find the hexadecimal equivalent of 157.

One of the useful things about hexadecimal is that it translates easily with binary. If you break up a binary number into four-bit groups (called *nibbles* i.e. small bytes), these little groups can individually be translated into one 'hex' digit.

Example 1.4 Convert 01101001 into hex. Divide the number into nibbles: 0110 and 1001. It is easy to see 0110 translates as $4 + 2 = 6$ and 1001 is $8 + 1 = 9$. So the 8 bit number is **69** in hexadecimal. As you can see, this is much more straightforward than with decimal, which is why hexadecimal is more commonly used.

Exercise 1.5 Convert 11101010 into a hexadecimal number.

An 8 bit system

The PIC is an 8 bit system, so it deals with numbers 8 bits long. The number 11111111 is the largest 8 bit number and equals 255 in decimal and FF in hex (work it out!). With PIC programming, different notations are used to specify different numbering systems (the decimal number 11111111 is very different from the binary number 11111111)! A binary number is shown like this: **b'00101000'**, a decimal number like this: **d'72'** , or like this: **.72** (it looks like 72 hundredths but it can be a lot quicker to write if you use decimal numbers a lot). The hexadecimal numbering system is default, but for clarity write a small h after the number (the computer will still understand it and it reminds *you* that the number is in hex) e.g. 28**h**.

When dealing with the inputs and outputs of a PIC binary is always used, with each input or output pin corresponding to a particular bit. A **1** corresponds to what is known as *logic 1*, meaning the pin of the PIC is at the supply voltage (e.g. +5V). A 0 shows that pin is at *logic 0*, or 0V. When used as inputs, the boundary between reading a logic 0 and a logic 1 is half of the supply voltage (e.g. +2.5V).

Finally, if at any stage you wish to look up what a particular instruction means, refer to Appendix C which lists of all of them with their functions.

Initial steps

The basic process in developing a PIC consists of five steps:

1. **Select** a PIC, and construct a program **flowchart**.
2. **Write** program (using Notepad provided with Microsoft Windows, or some other suitable development software).
3. **Assemble** program (changes what you've written into something a PIC will understand).
4. **Simulate** or **emulate** the program to see whether or not it works.
5. **'Blow'** or **'fuse'** the PIC. This feeds what you've written into the actual PIC.

Let's look at some of these in more detail.

Choosing your PIC

Before actually beginning to write the program, it is a very good idea to perform some preliminary tasks. First you need some sort of project brief – what are you actually going to make and what exactly must it do. The next step is to draw a circuit diagram, looking in particular at the PIC's inputs and outputs. Each PIC has a specific number of inputs and outputs, you should use this as one of the deciding factors on which PIC you should use and thus should make a list of all the inputs and outputs required. The PIC54 has up to 12 input/output pins (i.e. they have 12 pins which can be used as inputs *or* outputs), and the PIC55 has up to 20.

Example 1.5 The brief is 'design a device to count the number of times a push button is pressed and display the value on a single seven-segment display. When the value reaches nine it resets.'

1. The seven-segment display requires **seven** outputs.
2. The push button requires **one** input, creating a total of 8 input/output pins. In this case a PIC54 would therefore be used (see Figure 1.2).

Make sure you employ **strobing** where possible. This is particularly useful when using more than one seven-segment display, or when having to test many buttons. Example 1.6 demonstrates it best:

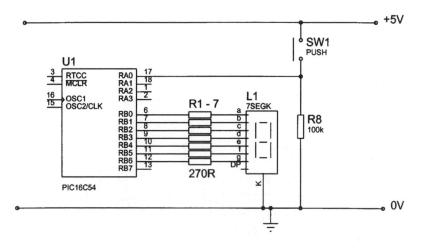

Figure 1.2

Example 1.6 The brief is 'to design a system to test 16 push buttons and display the number of the button pressed (e.g. button number 11) on two seven-segment displays'.

It would first appear that quite a few inputs and outputs are necessary:

1. The two seven-segment displays require seven outputs each, thus a total of **14**.
2. The push buttons require one input each. Creating a total of **16**.

The overall total is therefore 30 input/output pins, which exceeds the maximum for the PIC55. There are bigger PICs with more than 30 pins, however it would be unnecessary to use them as this value can be cut significantly.

By strobing the buttons, they can all be read using only 8 pins, and the two seven-segment displays controlled by only 9. This creates a total of 17 input/output (or I/O) pins, which is under 20. Figure 1.3 shows how it is done.

By making the pin labelled RC0 logic 1 (+5V) and RC1 to RC3 logic 0 (0V), switches 13 to 16 are enabled. They can then be tested individually by examining pins RC4 to RC7. Thus by making RC0 to RC3 logic 1 one by one, all the buttons can be examined individually.

Strobing seven-segment displays basically involves displaying a number on one display for a short while, and then turning that display off while you display another number on another display. RB0 to RB6 contain the seven-segment code for both displays, and by making RA0 or RA1 logic 1, you can turn the

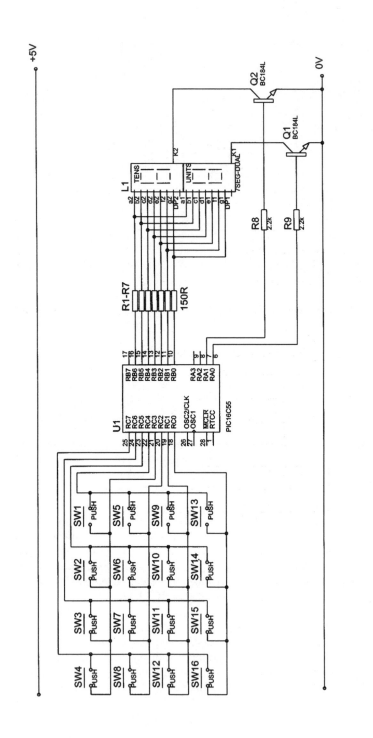

Figure 1.3

individual displays on. So the displays are in fact flashing on and off at high speed, giving the impression that they are constantly on. The programming requirements of such a setup will be examined at a later stage.

Exercise 1.6 Work out which PIC (54 or 55) you would use for a device which would count the number of times a push button has been pressed and display the value on four seven-segment displays (i.e. will count up to 9999).

After you have worked out how many I/O pins you will need, and thus selected a particular PIC, the next step is to create a program flowchart (Example 1.7). This basically forms the backbone of a program, and it is much easier to write a program from a flowchart than from scratch.

A flowchart should show the fundamental steps that the PIC must perform, showing a clear program structure. A program can have *jumps*, whereby as the PIC is stepping through the program line by line, rather than executing the next instruction, it jumps to another part of the program. All programs require some sort of jump, as all programs must loop – they cannot just end.

Example 1.7 The flowchart for a program to simply keep an LED turned on.

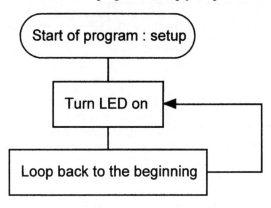

Figure 1.4

The setup box represents some steps which must be taken as part of the start of every program, in order to set up various functions – this will be examined later. Rectangles with rounded corners should be used for start and finish boxes, and diamond shaped ones for decisions.

Conditional jumps can also be used: *if* something happens, *then* jump somewhere.

Example 1.8 The flowchart for a program to turn an LED on *if* a button is pressed.

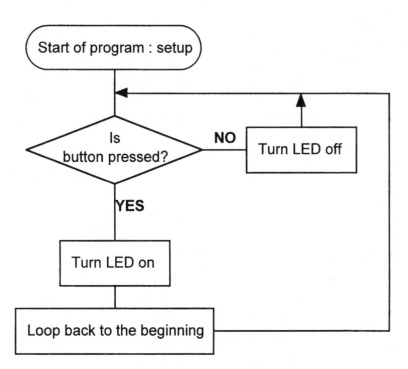

Figure 1.5

Sometimes a flowchart box may represent only one instruction, but sometimes it may represent a great deal, and such a diagram allows you to *visualize* the structure of your program without getting bogged down with all the nitty gritty instructions. Writing a program from a flowchart merely involves writing the instructions to perform the tasks dictated by each box, and in this way a potentially large program is broken down into bite-sized chunks.

Exercise 1.7 Draw the flowchart to represent the program required to make an LED flash on and off every second, and a buzzer to sound for one second every five seconds.

Writing

Once the flowchart is complete, you should load up a PIC program template on your computer (soon you will be shown how to create a sample template) and write your program on it. All this is done with a simple text program called Notepad which comes with Microsoft Windows (or another suitable development package).

Assembling

When you have finished writing your program, it is ready to be *assembled*. This converts what you've written (consisting mostly of words) into a series of numbers which the computer understands and will be able to use to finally 'blow' the PIC. This new program consisting solely of numbers is called the *hex code* or *hex file* – a hex file will have **.hex** after its name. Basically, the 'complicated' PIC language that you will soon learn is simply there to make program writing easier; all a *raw* program consists of is numbers (some people actually write programs using just numbers but this is definitely *not* advisable as it is a nightmare to fix should problems arise). So the *assembler*, a piece of software which comes with the PICSTART or MPLab package – called MPASM (DOS version) or WinASM (Windows version) – translates your words into numbers. If, however, it fails to recognize one of your 'words' then it will register an **error** – things which are *definitely* wrong. It may register a **warning** which is something which is *probably* wrong (i.e. definitely unusual but not necessarily incorrect). The only other thing it may give you is a **message** – something which *isn't* wrong, but shows it has had to 'think' a little bit more than usual when 'translating' that particular line. Don't worry if you are still a little confused by assembling, as all this will be revised as you go through the process of actually assembling your program.

Once you have assembled your program into a series of numbers, they get fused into ROM (Read Only Memory) of the PIC when you 'blow' the PIC, and there they stay forever, unless you have an erasable PIC (more on these later).

You should now be ready to begin writing your first program ...

The file registers

The key to the PIC are its file registers. Basically, if you understand these you're half way there. Imagine the PIC as a filing cabinet, with many drawers, each containing an 8 bit number (a byte). These drawers are the file registers. As well as these file registers there is the *working register*. This register is different because it is not part of the filing cabinet. It is needed because only one drawer (i.e. file register) may be open at one time. So imagine transferring a number from one drawer to another. First, you open the first drawer, take the number out then close it, now ... where is the number? The answer is that it is in the working register, a sort of bridge between the two file registers (think of it as the poor chap who has to stand in front of the filing cabinet). The number is temporarily held there until the second drawer is opened, upon which it is put away.

As you can see, each file register is assigned a particular number. You should call the file registers by their actual name when writing your program (as it is *much* easier to follow), and then the assembler will translate your names back to numbers when creating the hex file.

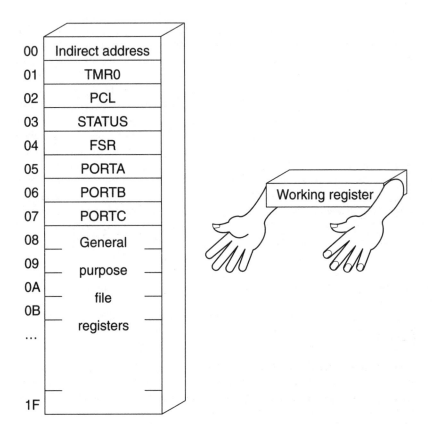

00	Indirect address
01	TMR0
02	PCL
03	STATUS
04	FSR
05	PORTA
06	PORTB
07	PORTC
08	General
09	purpose
0A	file
0B	registers
...	
1F	

Working register

Figure 1.6

Do not worry about the names or functions of these file registers, they will be discussed later on. However, to summarize, registers 00 to 07 have specific functions, and registers 08 to 1F are *general purpose* file registers, which you have complete control over. You can use general purpose file registers to store numbers and can give them whatever name you want. Naturally you will need to tell the assembler how to translate your own particular names into numbers. For example, if you were to use file register 0C to store the number of hours that have passed, you would probably want to call it something like **Hours**. However, as the assembler is running through your program, it will not understand what you meant by 'Hours' unless you first *declare* it. You will be shown how and where to declare your file registers shortly, when we look at a program template.

Before this, a brief introduction to registers 05 to 07 is required ...

The *ports* are the connections between the PIC and the outside world, its inputs and its outputs. The first port, Port A, has only 4 bits, i.e. it holds a nibble rather than a full byte and is the only register that does so. Each bit corresponds to a particular I/O (input/output) pin, so bit 0 of Port A corresponds to the pins labelled RA0 (pin 17 on the PIC54 and 6 on the PIC55). So when you write an 8 bit number into Port A, the four *most significant* bits are ignored, and likewise when you read an 8 bit number from Port A, the four most significant bits are read as 0.

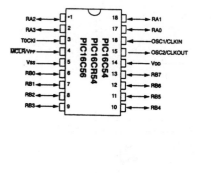 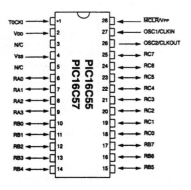

Figure 1.7

For example, let us say that RA0, RA1, RA2 and RA3 are acting as inputs and there is a push button between each input and +5V. If these push buttons are all pressed, the decimal number 15 (binary number 1111) would be in Port A. Conversely, if they are acting as outputs and are all connected to LEDs which were tied down to 0V, moving the number 15 into Port A would turn all four LEDs on.

Exercise 1.8 Considering the arrangement just mentioned, in order to create a chase of the four LEDs (see Figure 1.8), a series of numbers will have to be moved into Port A one after another. What will these numbers be (answers in binary, decimal or hexadecimal)?

Port B and Port C are simply other input/output ports just like Port A in all respects except that they have 8 bits (i.e. hold a byte). However, the PIC54 doesn't have a Port C, and so when using it, file register number 07 is another general purpose file register. It is recommended that you do *not* use it as a general purpose file register (unless you use up all the others) in order to maintain compatibility with the PIC55. In other words, if for some reason you want to run your program in a PIC55 chip, using Port C as a general purpose file register may foul things up.

LEDs

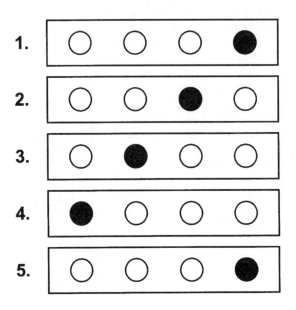

Figure 1.8

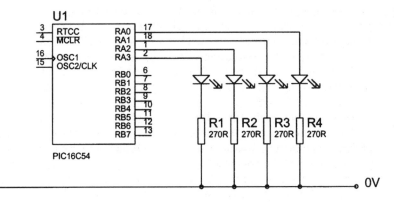

Figure 1.9

A program template

In this and subsequent sections you will begin to look at instructions. You may well find them confusing, but fortunately there are a few general rules you can use to decipher an unknown instruction. Firstly wherever you come across the letter **f** in an instruction, it refers to a **f**ile register. A **w** will nearly always mean **w**orking register, and a **b** stands for **b**it in the vast majority of cases. Finally, an **l** will usually stand for **l**iteral, which effectively means number. An instruction containing an **l** will therefore require a number to be specified afterwards.

Example 1.9
(Label)   bsf   porta, 0   ; turns on LED

There are a few fundamental elements to writing PIC program, one of these is line structure. Example 1.9 shows a sample line of programming. Optional first is a label which is useful if you want to jump to this place in the program. Then comes the actual instruction: **bsf** i.e. what are you *doing*. Third comes what are you doing it *to* (**porta, 0**), and lastly an explanation in your own words of what you have just done. It is important to note that you can write whatever you want in a PIC program as long as it is *after* a semicolon. Otherwise the assembler will try and translate what you've written (e.g. turns on LED) and will naturally fail and give you an **ERROR**. As the assembler scans through line by line, it will jump down to the next line once it comes to a semicolon.

I cannot stress how important it is to *explain* every line you write. First, what you've written may make sense as you write it, but there is a good chance that when you come back to it after a while, it will be difficult to understand. Second, it allows another person to read through your program with reasonable ease. It can sometimes be quite difficult to write a *good* explanation, as it should be very clear yet not too long. Don't get into the habit of basically copying out an instruction definition as your explanation, as shown in Example 1.10.

Example 1.10
**   bsf   porta, 0   ; sets bit 0 of Port A**

As this means very little at all (it is easy to see that bit 0 is being set), it is far better to say *why* you have written what you have, and what are its implications (as shown in Example 1.9).

Now let's look at a program template, bear in mind this is simply an example and you may want to add or remove headings for your own personal template. In general, with your whole program, it is a good idea to space things out, and divide relevant sections up with lines. I suggest creating these with equal signs (=), of course you need a semicolon at the start of such a line.

Program template

```
;************************************
; written by:                      *
; date:                            *
; version:                         *
; file saved as:                   *
; for PIC...                       *
; clock frequency:                 *

;************************************

; PROGRAM FUNCTION:——————————————
;————————————————————————————————

        list      P=16C5x
        include   "c:\pic\p16c5x.inc"

;============
; Declarations:

        porta     equ    05
        portb     equ    06
        (portc    equ    07)

        org       1FFh
        goto      Start
        org       0

;===========
; Subroutines:

Init    clrf      porta          ; resets input/output ports
        clrf      portb
        (clrf     portc)
        movlw     b'xxxx'        ; sets up which pins are inputs and which
        tris      porta        ;    are outputs
        movlw     b'xxxxxxxx'
        tris      portb
        (movlw    b'xxxxxxxx'
        tris      portc)
        retlw     0

;=============
```

```
; Program Start:
Start
          call      Init
Main
    (Write your program here)
END
```

In the little box made up out of asterisks (purely there to make it look nice), there are a couple of headings which allow another reader to quickly get an idea of your program. Where it has: **for PIC...** , insert a number such as 54 or 55 (making it: **for PIC54**), depending on which PIC you are using.

The clock frequency shows the frequency of the oscillator (*resistor/capacitor* or *crystal*) that you have connected to the PIC. The PIC needs a steady signal to tell it when to move on to the next instruction (in fact it performs an instruction every *four* clock cycles), so if, for example, you have connected a 4 MHz oscillator to the PIC (the fine details of such a setup are explained later) – i.e. four million signals per second – the PIC will execute one million instructions per second. The clock frequency would in this case be 4 MHz.

Much more important than these headings are the actual preliminary *actions* that must be performed. The line: **list   P=16C5x** is incomplete. Replace the **5x** with the number PIC you are using (e.g. 54), so a sample line would be: **list P=16C54**. This tells the assembler which PIC you are using.

The line: **include   "c:\pic\p16c5x.inc"** enables the assembler to load what is known as a *look-up* file. This is like a translator dictionary for the assembler. The assembler will understand most of the terms you write, but it may need to *look up* the translations of others. All the file registers with specific functions (00 to 07) are declared in the look-up file. When you install PIC software it will automatically create this file and put in a directory. I have suggested you rename it, calling the folder 'pic' instead, but naturally you can call it what you want. However you must show the path to get to the look-up file (e.g. c:\pic\p16c5x.inc).

Next comes the space for you to make your *declarations*. These are, in a sense, your additions to the translator dictionary. If you were to declare **Hours** as file register **0C**, you would write the following:

```
;============
; Declarations:
          Hours      equ      0Ch
```

You may also want to *re-declare* certain file registers with specific functions. This is because the assembler is sensitive to whether something is upper case or in lower case. For example, the look-up file declares file register 05 as **PORTA**. Personally I prefer writing it as **porta**, because it is quicker (I understand you may be happy to leave it as PORTA, but this example demonstrates the principle), so I will re-declare 05 as **porta** along with my other declarations:

;=============

; Declarations:

porta	**equ**	**05h**
Hours	**equ**	**0Ch**

This means I can write **porta** *or* **PORTA** and the assembler with understand *both* as file register 05. I also suggest declaring in order of increasing file register number.

Below the declarations are three lines which ensure the chip runs the program starting from the section labelled **start**. To understand this principle you must understand that every *instruction* line (i.e. not just a space or a line with some comments) has particular number (or *address*) assigned to it.

Example1.11

```
      start
0043       bsf    porta, 0   ; turns on LED
                             ; (This is to prove comments aren't counted)
0044       goto   start      ; loops back to start
```

Notice how only the lines with instructions have addresses (**start** is merely a label and not an instruction). The PIC54 and 55 will always start the program from the *last* address, and both have a maximum of 200**h** (512 in decimal) addresses (i.e. the size of their ROM), so 1FF**h** (511 in decimal) is the last address (remember the numbering starts from 0). Now, the allocation of addresses is systematic – counting up as you go down the program – *unless* you tell it otherwise. You can actually label the next line with a particular address, and then the ones which follow will continue counting up from there. This is done with the instruction **org**, followed by the address number you wish to give the next line.

Example 1.12

```
      start
0043       bsf    porta, 0   ; turns on LED
           org    3          ; makes the address number of the next
                             ;   instruction 3
0003       bsf    porta, 1   ; turns on buzzer
0004       goto   start      ; loops back to start
```

Notice how the instruction **org** is *not* given an address. Example 1.12 however would never work, because after executing address 0043, the chip would attempt to execute address 0044, but regardless it demonstrates the principle of the **org** instruction. In the template, **org** is used to label the instruction **goto   start** as address 1FF (the *last* instruction and thus the *first* to be executed). However subsequent instructions must start counting from 0, and thus the next instruction is: **org   0**. Writing the address by the instructions shows how it works:

```
            org      1FF
01FF        goto     start
            org      0


;============
; Subroutines:

0000   Init    clrf    porta    ;
0001           clrf    portb    ;
etc.
```

The first instruction to be executed (**goto   start**) makes the chip **go to** (*jump*) to the part of the program labelled **start**, and thus the PIC will begin running the program from where you have written **start**.

The next section of the template holds the *subroutines*. These are quite complicated and will be investigated at a later stage; all you need know at the moment is that the section labelled **Init** is a subroutine, and it is accessed using the **call** instruction. The actual subroutine **Init** should be used to set up all the particulars of the PIC. With the PIC5x series of chips, this mainly involves selecting which pins of the PIC are to act as inputs, and which as outputs. In other cases with more complex PICs, more setting up will be required. Please note that this setting up is put in the **Init** subroutine only to get it out of the way of the main body of program and thus hopefully make it neater and more reader friendly. First we use the instruction:

> **clrf   FileReg   ;**

This **clears** (makes zero) the number in a file register. We use it at the start of the setup subroutine to make sure the ports are reset at the start of the program. This is because when the PIC is reset, the states of the outputs are the same as they were before the PIC was reset. However in some cases where you want the states of the ports to be retained from the before the PIC reset, these clearing instructions may need to be removed. If the PIC doesn't contain a Port C, do not bother clearing it.

The next instruction is:

    movlw   number   ;

It **mov**es the literal (the **number** which follows the instruction – in the first case **b'xxxx'**) into the working register. Then the instruction **tris** takes the number in the working register and uses it to select which bits of the port are to act as inputs and which as outputs. A binary **1** will correspond to an input and a **0** corresponds to an output. Pins which you don't use are best made outputs.

Example 1.13 Using a PIC54, pins RA0, RA1 and RA3 are connected to push buttons. Pins RB0 to RB6 are connected to a seven segment display, and pins RA2 and RB7 are connected to buzzers. What should you write to correctly specify the I/O pins?

    movlw   **b'1011'**
    tris   **porta**
    movlw   **b'00000000'**
    tris   **portb**
    retlw   **0**

There are two things to notice: first there is no specification of Port C (naturally as the PIC54 doesn't have one), and second the reminder that bit numbering goes from right to left (it is easy to forget!).

Exercise 1.9
Using a PIC55, pins RA1 and RA2 drive LEDs, pins RA0 and RA3 are connected to temperature sensors, RB0 to RB6 control a separate chip, and RB7 is connected to a push button. RC1 to RC5 carry signals to the PIC from a computer, and all other pins are not connected. What should you write in the **Init** section of the program?

The instruction **retlw** is placed at the end of a subroutine, normally with a **0** after it.

    Finally the last part of the template holds **Start**, where the program begins. Notice that the first thing that is done is setting up the ports' inputs and outputs. After the line **call   Init**, there is the heading **Main** after which you write your program. At the end of your program, you must write **END**.

2
Exploring the PIC5x series

Your first program

For this chapter (and subsequent ones) it is assumed you are sitting in front of a computer which has the application Notepad. Do not worry if you don't have any actual PIC software at the moment, as the programs you write now can be assembled later, when you do actually get some software.

The first thing you should do is copy out a program template onto Notepad; try to make it as neat and clear as possible. Save the file as **template.asm** and make sure you select **any file** as the file type. The **.asm** shows that the file is an *assembly source*, i.e. it is something to be assembled, which makes it recognizable to the assembler. To begin with we'll be using the PIC54, so make the necessary alterations on the template (from now on do *not* simply **Save**, but instead **Save As**, so the file **template.asm** remains unchanged). Call this new file **ledon.asm.**

The first program you will write will be very simple. It simply turns on an LED (and keeps it on indefinitely). This will simply use two instructions: **bsf** and **goto**.

The instruction **bsf** sets (i.e. makes 1), a particular bit in a file register. You therefore need to specify the file register and the bit after the instruction (what you are doing it *to*).

Example 2.1   **bsf   portb, 5   ; turns on buzzer**

portb is the file register, and **5** is the number of the bit being set. There is a comma between the file register and the bit.

You should already be familiar with the instruction **goto** (remember **goto   start** from the template?). It makes the PIC jump to the section of the program you have labelled **start**. Naturally you can name the place to which you want it to jump anything you want, but it is a good idea to make it relevant to what is going on in the program in that particular section. Be careful, however, not to give sections the same name as you give to general purpose file registers, otherwise the PIC will get confused.

The first step of writing a program is assigning inputs and outputs. For this device we simply need one output for the LED. This will be connected to RA0 (pin 17) of the PIC. The second step is the program flowchart shown in Figure 2.1.

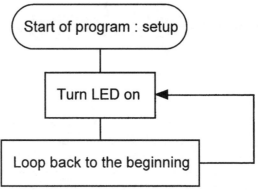

Figure 2.1

We can now write the program. You should be able to set up the inputs and outputs yourself (remember if a pin is not connected, make it an output). You can also have a go at writing the program yourself (it should consist of two lines).

The first box (Set up) is performed in the Init subroutine. The second box involves turning on the LED. This involves making RA0 high (+5V), and thus bit 0 of Port A should be 1 (i.e. *set*). To do this we use the instruction **bsf**. The line after ...

Main   **call**   **Init**   ;

... should therefore be:

      bsf   **porta, 0**   ; **turn on LED**

Remember, a program cannot just end; it must keep looping, so the next box involves making the program jump back to the beginning. The next line should therefore be:

      goto   **Main**   ; **loops back to Main**

Note that it should *not* go back to **Start**, as this will do the setting up all over again. Depending on how you wrote Port A in the program, you may need to redefine it in the declarations section. This would be necessary unless you wrote **PORTA** (i.e. in upper case).

The program is now ready to be assembled and you may want to check you have everything correct by looking at program in its entirety. This (along with all the other example programs) is shown in the program section in Chapter 7. This program has been given the name *Program A*. We now turn to assembling the program, but if you do not have any PIC software you will just have to read through the next section.

There are two main assembling packages, **MPASM**, and **MPASMWIN**. The first is a DOS based piece of software, and the latter is Windows based. If you double click on the MPASM icon, a screen will appear with seven parameters to be set by you. First, the assembler will ask you for the file which you want to assemble. Type the file (with its path) in the section labelled **Source File**. Then select the processor which the program you have written is for (P16C54, P16C71 etc.) in **Processor Type**. You then can select whether you want the assembler to produce a *cross reference file, listing file, hex dump file,* and/or an *object file*. The error file gives you a list of all the messages, warnings and errors which the assembler finds in your program, with the corresponding line numbers. This file isn't actually very useful, so you should select **no** for the **Error File** section. You shouldn't really need a cross reference file either, so for the **Cross Reference File** section, select **no** as well. The listing file is the useful one when it comes to tracking down your errors. It lists the errors actually with your program, beside where they occur. You are best off using the **Search** function in Notepad when in the listing file to find all the errors etc., in your program. So for the **Listing File** section, enter **yes**. Next on the list is the hex dump file – this is the file used by the testing packages and by the actual programmer which feeds the program into the chip. You should enter **inhx32** in the **Hex Dump Type** section. Finally you are asked whether or not you want an object file to be produced. This file contains the actual hexadecimal numbers which correspond to your program. If you have a special desire to see your file translated into numbers you can choose to have one produced, but there is no actual need to do so. When this is all ready, you may assemble (F10). You will then be told how many messages, warnings and errors have been found, and how many lines the assembler has gone through.

The Windows based assembler is basically the same, though Windows users will find it a lot more straightforward. You are given the same options (except in this case putting default in all cases will suffice). You are, however, given the option of making the assembler case sensitive or not (with the DOS version it is case sensitive). Everything else on the screen speaks for itself.

After you have sorted out all the errors etc., the assembler will produce the **.hex** file, which can be used to test program and later blow the final PIC.

Testing the program

There are basically three steps to trying out your program:

1. **Simulating**
2. **Emulating**
3. **Blowing** a PIC and putting it in a circuit

The first of these, **simulating**, is entirely software based. You simply see numbers changing on the computer screen and need to interpret this as to whether or not the program is working. The slight complication with the system involves inputs – it is quite easy for the computer to show you what is coming out, but how do you tell it the state of the PIC's inputs? You have to construct a file which tells the *simulator* (it comes with the PICSTART or MPLab package) the state of the inputs at particular times. Using this system you can get a fairly decent idea as to the success of your program.

Much more visual (and unfortunately much more expensive) is the **emulator**. There are various emulators: PICMASTER (from Microchip) and ICEPIC (from RF Solutions). For more information on the ICEPIC emulator consult Reference 1 in Appendix G. Emulators employ a probe in the shape of a PIC which plugs into your circuit board. You can then load your program and see on screen the state of all the PIC's file registers as you step through your program. The advantage of this is that the emulator responds to the actual inputs connected to the PIC on the circuit board, and the output changes are clearly visible, again on the circuit board.

The final step involves actually blowing a PIC with the program and putting it in a circuit; you should really only do this after having tested the program using one of the previous two methods. It is probably best to use a *UV erasable* PIC, whose program can be erased by exposure to ultraviolet light. These are more expensive than *one time programmable* PICs, and require some sort of UV exposure unit. However, they will turn out cheaper in the long run, because they can be used over and over again. When you program a normal PIC it stays that way, and if the program is wrong in any way you've just wasted a PIC. (With the PIC84 you can actually electronically erase the program, but more on this at a much later stage).

So for two of the three methods, you need some sort of circuit board to put your PIC into. As well as connecting the inputs and outputs to the correct pins of the PIC, there are other connections required. These are discussed in the hardware section which follows.

Hardware

Figure 1.7 shows the pin arrangements for the PIC54 and PIC55. You may already be familiar with these from a previous section. The pins labelled RA..., RB..., and RC... are I/O pins. VDD and VSS are the positive and 0V supplies, respectively. The positive supply should be between 2.5 and 6.25 V. The pin labelled T0CKI is the **T**imer **Z**ero **C**lock **I**nput, it receives signals and automatically counts them (if you set up the on board timer to do so). MCLR is the **M**aster **C**lear pin; basically it's a reset pin. The bar over the top means that it is *active low*. This means that when you make the pin low (0V), the chips drops what it's doing and returns to start. Figures 2.2 and 2.3 show possible ways of connecting it up.

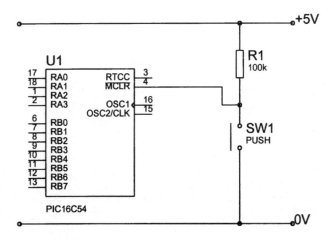

Figure 2.2

Figure 2.2 shows how to trigger the MCLR by means of a push button reset switch. The resistor is there because when the switch *isn't* pressed, the MCLR pin must be high and not simply disconnected, and with no resistor the positive and 0V supplies would be short circuited when the button is pressed.

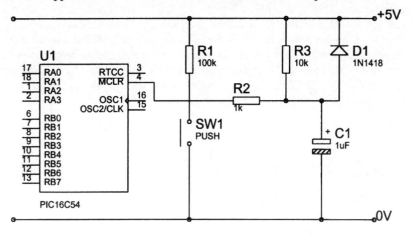

Figure 2.3

In most cases, however, you will want to use the circuit shown in Figure 2.3, since many power supplies take a short time to stabilize, and crystal oscillators also need a 'warm up' before the PIC actually starts. The circuit shown creates a small delay by keeping the MCLR low for a short period of time, and thus solves the problem.

Table 2.1

Cext	Rext	Average Fosc @ 5V, 25°C	
20 pF	3.3 k	4.973 MHz	± 27%
	5 k	3.82 MHz	± 21%
	10 k	2.22 MHz	± 21%
	100 k	262.15 kHz	± 31%
100 pF	3.3 k	1.63 MHz	± 13%
	5 k	1.19 MHz	± 13%
	10 k	684.64 kHz	± 18%
	100 k	71.56 kHz	± 25%
300 pF	3.3 k	660 kHz	± 10%
	5.0 k	484.1 kHz	± 14%
	10 k	267.63 kHz	± 15%
	160 k	29.44 kHz	± 19%

The chip also needs some sort of steady pulse to keep it going, such an oscillator can be created using a crystal or resistor/capacitor arrangement. The most reliable will probably be the crystal oscillator, as outside variables (such as temperature) don't effect it as much. Figures 2.4 and 2.5 show how such oscillators can be connected to the PIC. If you use a crystal, I recommend 2.4576 MHz for accurate timing, because though it doesn't look like it, 2.4576 is a much nicer number than 4 MHz for example, as you will soon see. Please note, however, that you could replace the crystal with a ceramic oscillator, which offers the same accuracy at a lower cost and smaller size. Table 2.1 shows the different frequencies generated by certain RC combinations.

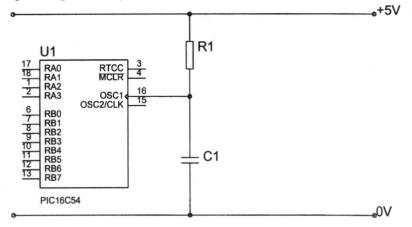

Figure 2.4

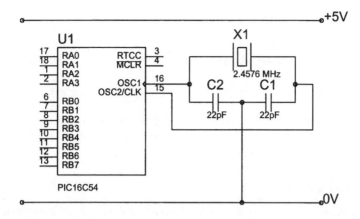

Figure 2.5

Simulating

You should now be ready to simulate your program. A piece of computer software uses the **.hex file** which the assembler has produced to allow you to step through your program line by line, or run through at full speed to see the program in action in real time. On earlier versions of PIC software, the simulator was called MPSIM, but on the more recent MPLab package which has a load of PIC functions rolled into one there is a slightly more visual one. If you have a program which doesn't respond to any inputs changing, you can just load the program and watch the outputs change (represented on screen). In the vast majority of cases however, you will need to create a *stimulus file* (the .sti file). This will change the states of the outputs on the simulator at certain times (e.g. on the 430th step, it could move the binary number **b'00110101'** into Port B etc.). This file must be in a specific format as shown below (the exclamation marks (!) are the same as semicolons (;), in that the PIC ignores the line when it comes to one of them).

! Stimulus file for SAMPLE.ASM

STEP	RB7	RB6	RB5	RB3	RB2	RB1	RB0	RA3	RA2	! Pins
3	0	0	0	0	1	0	0	1	0	! LED + ! buzzer
5	0	0	0	0	0	0	1	0	1	!
65	0	0	0	0	0	0	0	0	0	! all off
67	0	0	0	0	0	1	1	1	1	! etc.
254	1	0	1	0	1	1	0	0	0	!

Note: The maximum number of steps which can be assigned in this way is 12. Both types of simulator will use such a stimulus file, but work in slightly different ways. The instructions on how to run or step through are in both cases quite clear and you should be able to find your way around either package.

Emulation

As there are different types of emulators, with their own particular quirks, these will not be discussed in this book, however see Reference 1 in Appendix G for information on the ICEPIC emulator.

Blowing a PIC

Before blowing a PIC you need to construct some sort of circuit board. For this you need a circuit diagram. The required circuit diagram is shown in Figure 2.6. A printed circuit board for this and any of the other projects may be ordered, for more information, see Appendix F.

I have chosen a resistor/capacitor oscillator as it is cheaper, and accurate timing is not required in this project. If you have a UV erasable PIC you can now blow it (if you only have a one time programmable PIC don't bother). Place it in the programming module you are provided with (don't forget PICs are static sensitive), and load MPLab. From the **PicstartPlus** menu, select **Enable Programmer**. Providing that you have connected the programming module correctly, you should be given various windows relating to the programming of the PIC. If the package cannot find the programmer, first ensure it is turned on, and then if it still doesn't work, go into the **Options** menu and go to **Programmer Options**.

There will be a window showing the program which MPLab currently has in its memory. Each set of numbers in this window corresponds to a unit of memory. If this unit contains **FFF**, or **3FFF** etc., it is empty. Any other number corresponds to a particular instruction. The other window sets up the details of how you wish to program the PIC. Select the type of PIC you wish to program in **Device**, the type of oscillator which you are going to connect to the PIC in **Oscillator**. You must tell the programmer whether or not you wish the watchdog timer (WDT) to be on or off – remember if it is on, the PIC will constantly be resetting, so only turn it on if you specifically need it.

The next box you have control over is the **Code Protect** setting. If you wish yourself (and others) to able to put a programmed PIC in a programmer module and be able to read the program inside it, turn the code protect *off*. If you wish the program in the PIC to be protected so that it can't be read, turn code protect *on*. Although not possible with the PIC5x series, the PIC71 will allow you to turn the **Power Up Timer** on or off. This creates a small delay upon powering up the PIC before the program begins to run, thus allowing things like the crystal or a power supply to stabilize. On the P12C508 (the 8 pin PIC), you

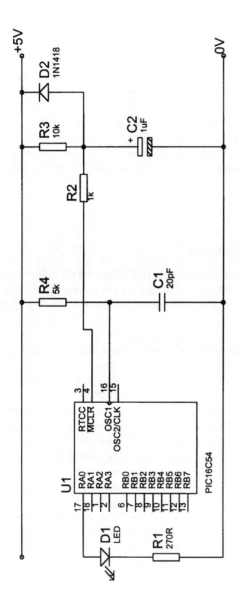

Figure 2.6

can select whether you want to use the MCLR pin provided as an MCLR pin, or as an input pin (all this is explained in the P12C50x section – Chapter 4).

The next box you need to sort out is the **Device ID**. You are best off clicking on the box, and selecting **Unprotected Checksum**.

All the settings are now correct, and you are ready to program your PIC. If you are programming an erasable PIC, click on **Blank**, to ensure the device is completely erased. once this is established you should go to the **File** menu, and select **Import**, and then **Download to Memory**. Find the **.hex** file of your program which the assembler produced and click OK. You should now see your program in numbers in the program memory window.

The big moment has finally arrived, you can now **Program** your PIC. A window will appear showing the stages in the programming, and then finally appear with a message: **Success** (hopefully), or possibly **Failure** (if something went wrong, e.g. PIC wasn't blank). You can now remove the PIC and place it in your board ... good luck!

All this just to see an LED turn on may seem a bit of an anticlimax, but there are greater things to come!

Using the testing instructions

A far more useful program would turn on an LED *if* a push button is pressed, and then turn it off when it is released. This will involve *testing* the state of the input pin connected to the push button. There are two basic methods of testing inputs:

1. Testing a particular bit in the port, using the **btfss** or **btfsc** instructions.
2. Using the entire **number** held in the port's file register to look at all the inputs as a whole.

In most cases you tend to test particular bits, and as there is only one push button, only one bit will need to be tested. The push button will be connected to pin RB0, and again the PIC54 will be used. Two I/O pins will be needed in this new device, and the flowchart is shown in Figure 2.7. The circuit diagram is shown in Figure 2.8.

Again, you should be familiar with the *set up* part, and be able to write it yourself. The next box requires the use of the new instruction **btfss**. This instruction tests a **bit** of a **file** register and will **skip** the next instruction if the bit is set (i.e. if it is high or logic 1). Its 'sister' instruction is **btfsc** which again tests a **bit** of a **file**, but this time skips the next instruction if the bit is **clear** (i.e. if it is low or logic 0). So to test the push button, the instruction line is:

btfss　　portb, 0　　; tests the push button

If the button pulls the input pin *high* when it is pressed, the PIC will execute the next instruction if the button is *not* pressed. In such a case the LED should be turned off and then the PIC should loop back to **Main**. The way to do this is to make the PIC *jump* to a section labelled something like **LEDoff**. This requires the instruction:

goto   LEDoff   ; jumps to the section labelled LEDoff

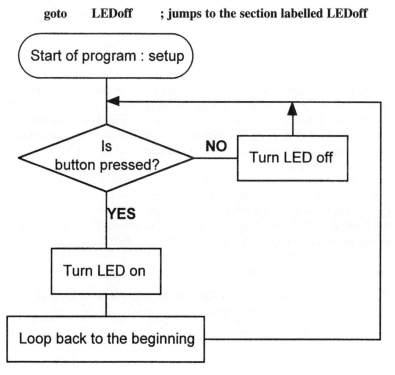

Figure 2.7

After this line is the instruction that will be executed if and only if the push button is pressed. This should therefore make the LED turn on. You should already know how do this, as well as the instruction that follows it which makes the PIC loop back to **Main**. This leaves us with the section labelled LEDoff. In this section the LED should be turned off, and then the PIC should loop back to **Main**. To turn a bit off use the instruction **bcf**. This clears a bit of a file register and works just like **bsf**. The next line is:

LEDoff   bcf   porta, 0   ; turns off LED

We finally come to the last instruction which again should make the PIC loop back to **Main**. You should be able to do this yourself. The program is now ready to be assembled, but again you may check that the program is correct by looking at whole program (named *Program B*). Go through the testing process again and you should be able to get it working. The result should be more satisfying (marginally), especially after all that effort. However with a bit of lateral thinking you can actually shrink the entire seven line program to something consisting only of three! You may be wondering how this can be, as we went through all the development processes and constructed a logical flowchart, but somehow there is a much better way.

Sometimes it helps to step back from the problem and look at it in a different light. Instead of looking at the button and LED as separate bits in the two ports, let's look at them with respect to how they effect the entire number in the ports. When the push button is pressed the number in Port B is **b'00000001'**, and in this case we want the LED to turn on (i.e. make the number in Port A **b'00000001'**). When the push button isn't pressed, Port B is **b'00000000'** and thus we want Port A to be **b'00000000'**. So instead of testing using the individual bits we are going to use the entire number held in the file register (think back to the two different testing methods introduced at the start of this section). The entire program merely involves moving the number that is in Port B into Port A. As you know this cannot be done directly and involves the moving of the number in Port B to the working register, and then moving the number from the working register into Port A. To move (in fact *copy*) the number from Port B into the working register we need the following instruction:

> **movf   FileReg, w   ;**

This **mov**es the number from a **f**ile register into the **w**orking register. This instruction is very often abbreviated to:

> **movfw   FileReg   ;**

This instruction will do exactly the same thing, and is translated to the same number by the assembler. So the instruction to move the number from Port B into the working register is:

> **movfw   portb   ; moves the number in Port B to the**
> **   ;   working reg.**

Then to move the number into Port A, we need the instruction:

> **movwf   FileReg   ;**

This **mov**es the number from the **w**orking register into a **f**ile register. To move the number from the working register into Port A we would write:

> **movwf   porta   ; moves the number from the w. reg. into**
> **   ;   Port A**

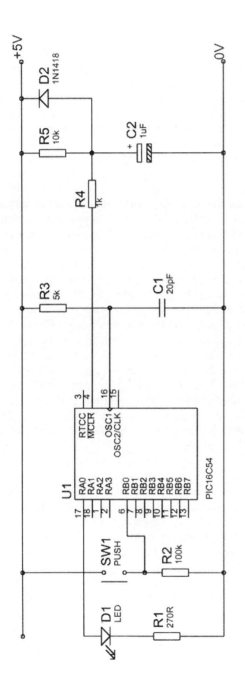

Figure 2.8

After these two lines we need only loop back to **Main** so it cycles through these two lines constantly. Please note this shorter technique can only be used because the push button and LED are connected to the particular pins described in this example. Unless you specifically connect them up so that the technique works, it is unlikely to do so. This shorter program is shown as Program C in Chapter 7.

The circuit diagram for this project is the same as with the previous version, which is shown in Figure 2.8. The next section will introduce timing which is where the PIC will really begin to get useful.

Timing

The PIC comes with an on board timer called **TMR0** (in more advanced chips there is more than one timer, e.g. TMR1, TMR2 etc.). As you may remember, TMR0 (said timer zero) is file register number 01. It has two basic modes: counting an *internal* or *external* signal. When on the internal counting mode, the number it holds counts up at a constant rate (depending on the oscillator attached to the PIC). When counting external signals, it counts the number of signals received by the timer zero clock input (pin 3 on the PIC54 and 1 on the PIC55). When the number passes 255, it resets and continues from 0 again, as with any file register (this is called *rolling over*). As you can already see, there are various settings for the TMR0 and these can be controlled by the bits in the **OPTION** register. This register will *not* be familiar as it wasn't on the diagram showing the file registers (Figure 1.6). This is because it isn't a file register that you can directly access (at least on the PIC5x series). In order to put a number into it, you first load the number into the working register, and then write the instruction: **option**. This automatically takes the number from the working register and moves it into the OPTION file register. The bits in the OPTION register are allocated as shown on page 35.

This may be hard to follow, but this is basically how all file registers are explained in the PIC databook, so it is important to be familiar with the format. In the OPTION register each bit controls a particular setting, except bits 6 and 7. As you can see they have no purpose and are read as 0. Bit 5 (RTS) is the *TMR0 signal source*, and defines whether TMR0 is counting internally (using the PIC's oscillator) or externally (counting signals on the T0CKI pin). Bit 4 (RTE) is the *TMR0 signal increment*, and is fairly irrelevant if counting internally, but can be important if counting external signals. It selects whether TMR0 counts up every time a signal drops from logic 1 to logic 0 (i.e. *falling edge triggered*), or when the signal rises from logic 0 to logic 1 (i.e. *rising edge triggered*).

Bit no.	7	6	5	4	3	2	1	0	
	-	-	RTS	RTE	PSA	PS2	PS1	PS0	Rate
	0	0				0	0	0	1:2
						0	0	1	1:4
						0	1	0	1:8
						0	1	1	1:16
						1	0	0	1:32
						1	0	1	1:64
						1	1	0	1:128
						1	1	1	1:256

Prescaler assignment
0 - if you want the prescaler to be used by the TMR0
1 - if you want prescaler to be used by the WDT

TMR0 signal increment
0 - if you want to count up when the signal drops
1 - if you want to count up when the signal rises

TMR0 signal source
0 - if you want to count an internal signal
1 - if you want to count an external signal

Now we come to the *prescaler* bits (PS2, PS1 and PS0). As you already know, the PIC divides the frequency of the oscillations it receives from its oscillator (crystal, R/C etc.) by four, and uses this as its driving frequency. This same value is used by TMR0 when counting internally. Let's take a typical oscillator frequency of 2.4576 MHz. This is divided by four leaving 0.6144 MHz, in other words a signal which oscillates 614 400 times a second. When trying to use TMR0 to count seconds, minutes and even days, it is clear that a file register which counts up so fast is of little use. TMR0 would have to count up to 614 400 for one second to pass, but of course it resets at 255 and would never reach this number. TMR0 has to be therefore *prescaled*, i.e. its frequency needs to be reduced. By the use of bits 0 to 2 in the OPTION register, TMR0 can automatically be prescaled by up to 256 times. When using TMR0 to count seconds and minutes etc., it would be necessary to prescale it by the maximum amount. Prescaling TMR0 by 256 divides the frequency of 614 400 Hz by 256, to 2400 Hz (surprisingly the numbers work out nicely!). So even with maximum prescaling, TMR0 still counts up once every 1/2400th of a second. We need to prescale it further ourselves, and this will be explained shortly.

The only bit left unexplained is bit 3 (PSA), the *prescaler assignment* bit. This introduces the idea of a *WDT* or *watch dog timer*, which is explained in a later section. Basically, without complicating things too much, this bit selects whether it is the WDT that is being prescaled or the TMR0 – you can only prescale one of them. Which ever one *isn't* being prescaled can still run, but with no reduction of the timer's frequency (i.e. a prescaling factor of 1).

Example 2.2 What number should be moved into the OPTION register in order to be able to use the TMR0 efficiently to eventually count the number of seconds which have passed?

Bits 6 and 7 are always 0.
TMR0 is counting *internally*, so bit 5 (RTS) is 0.
It's irrelevant whether TMR0 is *rising* or *falling edge triggered* so bit 4 (RTE) is 0 or 1, (let's say 0).
Prescaling for TMR0 is required, so bit 3 (PSA) is 0.
Maximum prescaling of 256 is required, so bits 2 to 0 (PS2-0) are all 1.

Hence the number to be moved into the OPTION register is: **00000111**.

Exercise 2.1 What number should be moved into the OPTION register in order to be able to use the TMR0 to count the number of times a push button is pressed?

Exercise 2.2 **Challenge!** What number should be moved into the OPTION register so that TMR0 can keep track of the number of times a push button is pressed, and reset when the maximum of 1023 presses is reached?

Now that you know what number to move into the OPTION register, you need to know *how* to move it. This calls for a familiar instruction: **movlw**. As you may remember, this moves the number that follows it into the working register. Then the instruction **option** moves the number from the working register into the OPTION register.

Example 2.3 **movlw   b'00000111'   ; sets up TMR0 to count**
            option   ;   internally, prescaled by 256

Notice how the explanation describes the *two* lines – rather than doing each one in turn, it makes sense to look at the instruction *pair*. As you are unlikely to want to keep changing the TMR0 settings it is a good idea to place the above instruction pair in the **Init** subroutine, to keep it out of the way.

If you want to be timing seconds and minutes, you need to perform some frequency dividing yourself. This is basically the same as prescaling, but as it takes place after the prescaling of TMR0, we should call it *postscaling*. This requires quite a complex instruction group, but let's try to build it up step by step. First, the essence of postscaling is counting the number of times a rising file register (like the TMR0) reaches a certain value. For example, we need to wait until the TMR0 counts up 2400 times, for one second to pass. This is the same as waiting until the TMR0 reaches 30, for a total of 80 times, because $30 \times 80 = 2400$ (think about it).

How do we know when TMR0 has reached 30? We subtract 30 from it, and see whether or not the result is zero. If TMR0 *is* 30, then when we subtract 30 from it, the result will be zero. However, by subtracting 30 from the TMR0 we are changing it quite drastically, so we use the command:

subwf   **FileReg, w**

This **sub**tracts the number in the working register from the number in a file register. The **,w** after the specified file register indicates that the result is to be placed back in the working register, thus leaving the original file register number *unchanged*. In this way we can subtract 30 from TMR0, without actually changing the number in TMR0, i.e. see what *would* happen to TMR0 if we were to subtract 30.

The next problem is finding out whether or not the result of the operation mentioned above is zero. This is done using one of the PIC's *flags* mentioned in Chapter 1. The flag we use is the *zero flag*. A flag is merely one bit in the *STATUS* file register (number 02), which is automatically set or cleared depending on certain conditions. The zero flag is set when the result of an operation is zero, and is cleared when the result isn't zero. You already know the instruction for testing a bit in a file register, in this case the instruction line would be:

btfss   **STATUS, Z**   **; tests the zero flag (skip if the result was 0)**

Rather than specifying the bit number after the file register, as is normally the case (e.g. **porta, 0**) – which in this case would be 2 – it is advisable to write **Z**, because it is understood by the assembler (with the help of a lookup file) and it is easier for you to understand. There are only a few select cases where this kind of substitution may be used.

So far, we have managed to work out when the TMR0 reaches the number 30. We need this to happen 80 times for one second to pass; this is best done using the following instruction line:

decfsz   **FileReg, f**

This will **dec**rement (subtract one from) a file register, and skip the next instruction if the result is zero. This is in effect a shortcut, and the identical operation could be performed over numerous steps, including the testing of the zero flag. Thus if the number in the specified file register is 80, the PIC will pass this line 80 times until it skips. If the next instruction is a looping instruction (i.e. one which makes the PIC jump back to the beginning of this timing section, the PIC will keep looping until the number in the file register reaches 0 (i.e. it will loop 80 times), after which it will skip the looping instruction and proceed onto the next part of the program. For this whole timing concept to work, the PIC must only execute this **decfsz** instruction when the TMR0 has advanced by 30 (e.g. gone from 0 to 30 *or* from 30 to 60 etc.). If we are in a looping system, it is all very well to test for TMR0 to reach 30 the first time round, but it will take another 256 advances of TMR0 to reach 30 for a second time (the TMR0 will continue counting up past 30, reset at 255, and then continue from 0). We could therefore reset TMR0 every time it reaches 30, but other parts of the program may be using it and would be relying on it counting up steadily and continuously. A better solution is to change the number you are waiting for TMR0 to reach. The second time in the loop it would be necessary to test for TMR0 to reach 60 (i.e. 30 + 30), and then the next time 90 (60 + 30) etc. The number we are testing for should therefore be held in a file register (let's call it **Mark30**, because it **mark**s when TMR0 has advanced by **30**), and every time the TMR0 'catches up' with **Mark30**, 30 must be added to it. The instruction pair for this involves a new instruction:

    addwf   **FileReg, f**   **;**

This **add**s the number in the working register to the number in a file register, and leaves the result in the file register. So we need to move the number we want to add to the file register into the working register first. The required instruction pair to add the decimal number 30 to a file register called **Mark30** would therefore be:

    movlw   **d'30'**   **; adds 30 to Mark30**
    addwf   **Mark30, f**   **;**

When we need to access this number, it will be necessary to move (in fact *copy*) the number from the file register to the working register. As you know this involves the instruction **movfw.**

The file register which we are decrementing (which holds the number 80 to start with) shall be called **Post80** (Timer **Post**scaler by a factor of **80**).

The program section which follows is the entire instruction set required to create a one second delay. The first four lines where numbers are being moved into Mark30 and Post80 may be placed in the **Init** subroutine. Read through the instruction set carefully, we will be using this technique in the next example program. Please note that GPF stands for general purpose file register.

```
        movlw   d'30'          ; moves the decimal number 30 into
        movwf   Mark30         ;   the GPF called Mark30, the marker

        movlw   d'80'          ; moves the decimal number 80 into
        movwf   Post80         ;   the GPF called Post80, the first
                               ;   postscaler
TimeLoop
        movfw   Mark30         ; takes the number out of Mark30
        subwf   TMR0, w        ; subtracts this number from the
                               ;   number in TMR0, leaving the result
                               ;   in the working register (and leaving
                               ;   TMR0 unchanged)
        btfss   STATUS, Z      ; tests the zero flag - skip if set, i.e. if
                               ;   the result is zero it will skip the next
                               ;   instruction
        goto    TimeLoop       ; if the result isn't zero, it loops back to
                               ;   'Loop'

        movlw   d'30'          ; moves the decimal number 30 into the
                               ;   working
        addwf   Mark30, f      ;   register and then adds it to Mark30

        decfsz  Post30, f      ; decrements Post80, and skips the next
                               ;   instruction if the result is zero
        goto    TimeLoop       ; if the result isn't zero, it loops back to
                               ;   'Loop'

; When it reaches this point, 1 second has passed

        movlw   d'80'          ; resets Post80, moving the number 80
        movwf   Post80         ;   back into it
```

The next example project will be an LED which turns on and off every second and a buzzer which sounds for one second every five seconds. This will involve two outputs, one for the LED and one for the buzzer. The LED will be connected to RA0, and the buzzer to RB0. The oscillator should be accurate so a crystal arrangement will be used, running at 2.4576 MHz. The program flowchart for this project is shown in Figure 2.9, and the circuit diagram in Figure 2.10.

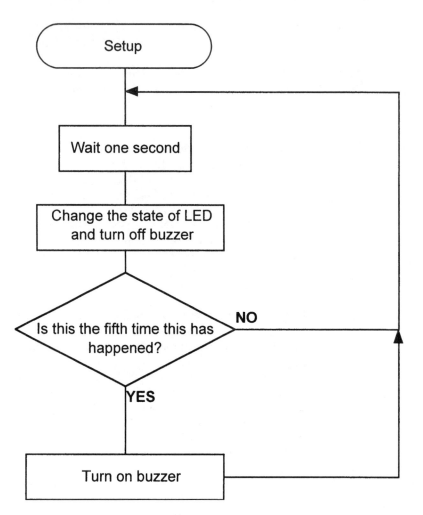

Figure 2.9

The setup should present no problems, remember to define any general purpose file registers such as Mark30 and Post80 using the **equ** instruction. You can make them file register numbers 08 and 09 for example. In the **Init** subroutine you may want to specify the number that goes in the OPTION register.

The instruction set for the whole of the box 'Wait one second' is the program section mentioned previously which creates a 1 second time delay. At the end of the section (the line after the **movwf** instruction), the state of the LED must be changed (if it is on, turn it off and vice versa). There are two methods of achieving this. First, the current state of the LED can be tested (using the **btfss**

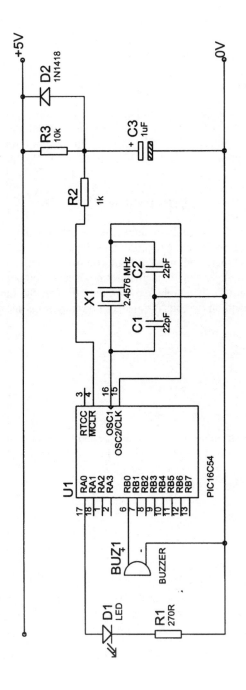

Figure 2.10

or **btfsc** instructions), after which the PIC branches off to one of two sections depending on the LED's state, which will then either turn it on or off. Far easier when the rest of the I/O port is empty (there are no other connections to Port A apart from the LED), is to use the following instruction:

> **comf**   **FileReg, f**   ;

This instruction **com**plements (toggles the state of all the bits in) a file register, and leaves the result in the file register. We can use this because even though it will effect all the other bits in Port A (RA1, RA2 and RA3), this doesn't matter as they aren't connected to anything. To toggle the state of the bits in Port A the instruction would be:

> **comf**   **porta, f**   ; toggles the state of the LED

However in most cases it won't be possible to simply toggle (change the state of) *all* the bits in a file register, so selective toggling must be carried out. This is done using the *exclusive OR* logic command. A logic command looks at one or more bits (as its inputs) and depending on their states produces an output bit (the result of the logic operation). The table showing the effect of the more common *inclusive OR* command on two bits (known as a *truth table*) is shown below.

Inputs		Result
0	0	0
0	1	1
1	0	1
1	1	1

The output bit (**result**) is high if either the first **or** the second input bit is high. The exclusive OR is different in that if *both* inputs are high, the output is low:

Inputs		Result
0	0	0
0	1	1
1	0	1
1	1	0

One of the useful effects is that if the second bit is 1, the first bit is toggled, and if the second bit is 0, the first bit isn't toggled (see for yourself in the table). In this way certain bits can selectively be toggled. If we just wanted to toggle bit 0 of a file register, we would exclusive OR the file register with the number **00000001**. This is done using one of the following instructions:

> **xorwf**   **FileReg, f**   ;

This exclusive **OR**s the number in the working register with the number in a file register, and leaves the result in the file register. Each bit is exclusive ORed to each other according to bit number (bit 0 with bit 0, bit 1 to bit 1 etc.). Alternatively it may be more suitable to use:

xorlw   number   ;

This exclusive **OR**s the number in the working register with a literal (**number**).

Exercise 2.3 Two instructions are needed to toggle bits 3, 5, and 7 of Port B, what are these two lines?

The other task that must be completed is turning off the buzzer. Most of the time the buzzer won't have been on anyway, but for the one in five times that it *is* on, this will turn it off after one second has passed. This is done using the **bcf** instruction.

Finally we need to see if this is the fifth time one second has passed (i.e. has five seconds passed?). This is done, as before, using the **decfsz** instruction. Use another general purpose file register called **_5Second** (the underscore at the start of the name is there because a file register name cannot start with a number). The number 5 should be moved into it to begin with, and then after the instruction is reached five times, it will skip the next instruction, which should therefore be some sort of looping instruction. After the number reaches 0, and therefore five seconds have passed, the number 5 should be moved back into **_5Second**, because otherwise it will take another 256 seconds for the value to reach 0 again (as with the resetting of **Post80** in the previous example). When five seconds have passed, the buzzer should be turned on, and then program loops back to the beginning.

The whole program is shown in Program D in Chapter 7; again you should go through the testing processes and get the program to work.

So far we have covered quite a few instructions and it is important to keep track of all of them, so you have them at your fingertips. Even if you can't remember the exact instruction name (you can look these up in Appendix C), you should be familiar with what instructions are available.

Exercise 2.4 What do the following do? **bsf, bcf, btfss, btfsc, movlw, movwf, movfw, decfsz, comf, subwf, addwf, equ, option, goto, tris, iorlw, iorwf, xorlw** and **xorwf**. (Answers in Appendix C).

Explain also the significance of **,f** or **,w** after the specified file register, with certain instructions, such as **subwf, addwf, comf,** and **decfsz** etc. (Answers in Appendix H).

There will also be another example project using most of the ideas we have so far covered: a traffic lights system. There will be a set of traffic lights for motorists (green, amber and red), and a set of lights for pedestrians (red and green), with a button for them to press when they want to cross. This makes a total of five outputs and one input, and thus the PIC54 will be used.

The green, amber and red motorists' lights (LEDs) will be connected to RB0, RB1, and RB2 respectively. The pedestrian push button shall go to RA0, with the green and red pedestrian lights to RB4 and RB5 respectively. The circuit diagram is shown in Figure 2.11.

To start with, the motorists' light should be green, with all the others off, until the push button is pressed. The red pedestrian light should be on, and the green one off. All this should present no great problem, however rather than setting and clearing the individual bits, simply move the correct number into Port B.

Exercise 2.5 What *two* lines will be used to get the LEDs in the correct states?

There then needs to be a loop where the pedestrian's push button is tested continually, the PIC should only jump out of the loop when the button is pressed.

Exercise 2.6 What *two* lines will this loop consist of?

As soon as the button is pressed (i.e. after the loop is jumped out of) the amber motorists' light should be turned on, and the green one turned off. There should be no change to the pedestrians' lights.

Exercise 2.7 What *two* lines will accomplish these required output changes?

As the flowchart in Figure 2.12 shows, there are quite a few time delays required, and rather than copy the same thing over and over again for each time delay, it makes sense to use a time delay *subroutine*. Subroutines are quite complicated things to understand fully and so we will merely *use* one this time and they will be properly explained in the oncoming section on seven-segment displays. All you need know (and you should already know this from studying the program template) is that when you access a subroutine, the PIC jumps to a certain place in the program, and when you are finished it will return to where it left off. To access a subroutine the instruction is **call**, and to return to the line after the call instruction you need to write **retlw**. This instruction must always be followed by a number, and in cases where this number is not important simply write **0** (as you may remember from the **Init** subroutine which you have grown to know and love). Basically we can write a subroutine to create a delay, *call* the subroutine when we want the delay to happen, and then know that the PIC will return after the delay to where it left off. To be able to use the delay subroutine for all delays, the delay would have to last for the smallest unit of time being used. In this case it is half a second, as this is the time between flashes when the 'green man' (green pedestrian light) is flashing. In the first case of the amber motorists' light, a delay of two whole seconds is required. The simple answer is to *call* the half second delay subroutine four times.

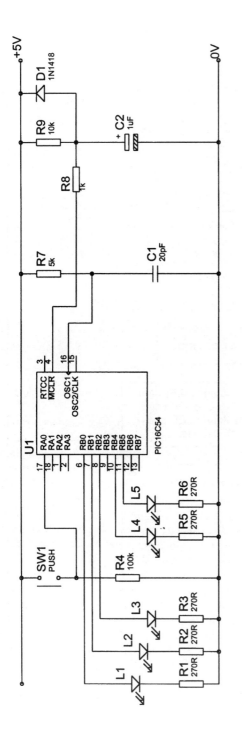

Figure 2.11

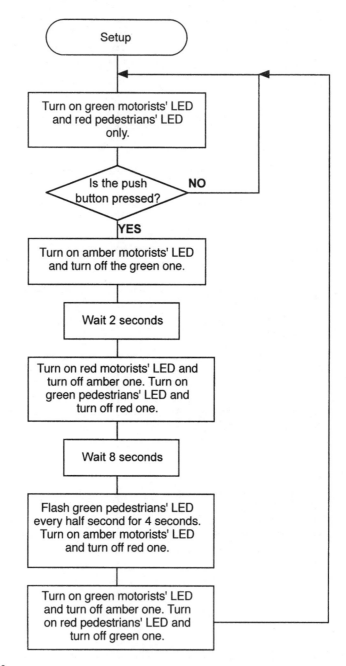

Figure 2.12

Exercise 2.8 Which *four* lines should be used to create a delay of two seconds? Do not bother writing the actual delay subroutine, merely give it a name.

After this two second delay the red motorist's light must be turned on, and the amber one off. The red pedestrian light must be turned off, and the green one on.

Exercise 2.9 Which *two* lines will make the required output changes.

Now an eight second delay is required, and this is solved by calling the half second delay subroutine 16 times. You could simply write the **call** instruction 16 times, however this would be a waste of lines and a shorter method is more desirable. We instead use a loop, using a general purpose file register to keep track of how many times the subroutine has been called.

```
          movlw    d'16'        ; moves the decimal number 16 into the
          movwf    Counter16    ;    general purpose file register called
                                ;    Counter16
Loop8     call     delay        ; creates half second delay
          decfsz   Counter16    ; does this sixteen times
          goto     Loop8        ; loops back until eight seconds have
                                ;    passed
```

We initially put the number 16 into the GP file register, and then decrement it every time the delay is called. When this has happened 16 times the PIC skips the **goto** instruction (remember the instruction **decfsz: dec**rement the file register and skip the next instruction if the result is zero).

After the eight seconds, the red motorists' light must be replaced for the amber one.

Exercise 2.10 What *two* lines will make the required change to the outputs?

Now the green pedestrian light must flash on and off every half second for four seconds (the output bit should thus toggle every half second, eight times). This is done in much the same way as with the 8 second wait: have a general purpose file register decrementing until it reaches 0.

Exercise 2.11 **Challenge!** What *seven* lines will this loop comprise of? Call the general purpose file register **Counter8** and move the correct number into it in the *two* lines before the loop.

The traffic lights now return to their original states, and the PIC can loop back to **Main**. All that remains is the time delay subroutine. This will be just like the one second time delay you have already met, except that this time the two postscaling values (previously 30 and 80) must multiply together to form 1200 (= 2400/2). Two such numbers are 15 and 80. This delay subroutine only differs from that previous instruction set in that one of the postscaling values is different and that the line ...

> **retlw   0   ;**

... must be added at the end.

Exercise 2.12 Write the time delay subroutine to wait half a second (don't forget to call it by the name you have been using in the main body of the program).

You have basically written this whole program yourself; to check the entire program as a whole, look at Program E in Chapter 7.

Seven-segment displays

Using seven-segment displays is quite a complicated task, but PIC allows you to display whatever you want. Obviously all the numbers can be displayed, but also most letters: A, b, c, C, d, E, F, h, H, i, I, J, l, L, n, o, O, P, r, S, t, u, U, and y.

The pins of the seven-segment display should all be connected to the same I/O port on the PIC, in any order (this may make PCB design easier). The spare bit may be used for the dot on the display. Make a note of which segments (a, b, c, etc.) are connected to which bits.

Example 2.4 Bit 7 = d, Bit 6 = a, Bit 5 = c, Bit 4 = g, Bit 3 = b, Bit 2 = f, and Bit 1 = e.

The number to be moved into Port B when something is to be displayed should be in the format **dacgbfe-** (it doesn't matter what bit 0 is as it isn't connected to the display), where each letter corresponds to the required state of the pin going to that particular segment.

The segments on a seven-segment display are labelled as shown in Figure 2.13.

So if you are using a common cathode display (i.e. make the segments high for them to turn on – see Figure 2.14), and you want to display (for example) the letter **P**, you would turn on segments a, b, e, f, and g.

Given the situation in Example 2.4, where the segments are arranged **dacgbfe-** along Port B, the number to be moved into Port B to display a **P** would be **01011110**. Bit 0 has been made 0, as it is not connected to the display.

Example 2.5 If the segments are arranged **dacgbfe-** along Port B, what number should be moved into Port B, to display the letter **I**, and the letter **C**?

The letter **I** requires only segments b and c (or e and f) so the number to be moved into Port B would be **00101000** or **00000110**.

The letter **C** requires segments a, d, e, and f, so the number to be moved into Port B would be **11000110**.

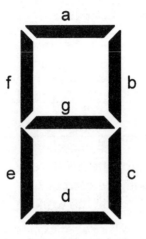

Figure 2.13

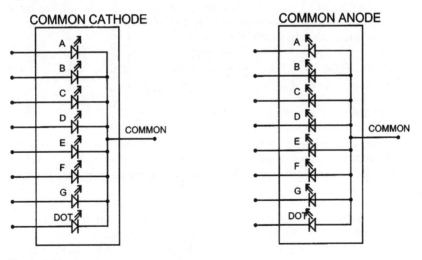

Figure 2.14

Exercise 2.13 If the segments are arranged **dacgbfe-** along Port B, what number should be moved into Port B to display the numbers 0, 1, 2, 3, 4, 5, 6 ,7, 8, 9, A, b, c, d, E, and F?

This conversion process can be carried out in various ways, but by far the simplest involves using a *subroutine*. To convert the number you want displayed into an actual display code, a decoding subroutine should be used. The general idea is that you first load the number to be displayed into the working register, then *call* (i.e. access) the subroutine, which will then return to the program with the appropriate code in the working register.

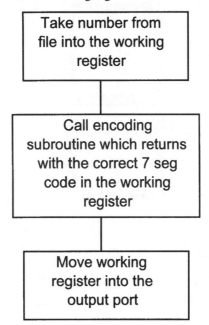

Figure 2.15

Let's call the subroutine **_7SegDisp**, and store the number we want displayed in a file register called **Display**. The seven-segment display will be connected to Port B. The instruction set in the main body of the program that would be required is:

movfw	**Display**	**; takes the number out of Display**
call	**_7SegDisp**	**; accesses the conversion subroutine**
movwf	**portb**	**; loads the correct code into Port B**

As you can see, nothing clever happens here. Where the actual conversion takes place is outside the main body of the program, in the subroutine. The subroutine uses the *program counter* (file register number 02). On the diagram showing the layout of file registers, this was given the name PCL – this stands for program counter lower. *The program counter stores the address of the next instruction to be executed.* There are 512 addresses in the PIC54 and 55, so clearly the program counter must be able to hold a number as large as 511 (remember one of the addresses is numbered 0). The PIC is an 8 bit system, so its file registers hold a maximum of 255, thus in order for the program counter to store a number as big as 511, it is split over two file registers, one holds the *lower* bits (0 to 7), and another holds the *upper* bits (8...). Despite the number being split over two file registers, the PCL is the only one we can actually change, but fortunately it is all we need to change in this case.

Example 2.6

0043   bsf   porta, 0   ; turns on LED

As the PIC is executing this line, the contents of the program counter would be **0044**, as the *next* instruction to be executed would be the one with address number **0044**.

Example 2.7

```
     start
0043     btfss portb, 0   ; tests push button
0044     goto  On
0045     goto  Off

0046 On  bsf   porta, 0    ; push button isn't pressed, so turn on LED
0047     goto  start       ; loop back to start

     Off
0048     bcf   porta, 0    ; push button is pressed, so turn off LED
0049     goto  start       ; loop back to start
```

This example shows some more complicated aspects of the program counter. To begin with, at address **0043**, a bit is being tested. If the bit is set, the number **0045** is moved into the program counter, if it is cleared the number **0044** is moved into the program counter. At address **0044**, the number **0046** is moved into the program counter, and at address **0045**, the number **0048** is moved into the program counter. I have deliberately written the **Off** label on the line above the instruction to demonstrate that it doesn't matter if the label isn't *physically* on the same line. Even though the label **Off** isn't *physically* on the same line as ...

bcf   porta, 0   ;

... it is still labelling that particular address, so the instruction:

goto   **Off**   **;**

... still moves the number **0048** into the program counter.

Exercise 2.14 In the above example, what number is being moved into the program counter at address **0047**, and what about at **0049**?

Is all very well to know what number is in the PC (program counter) at certain times, but it's far more useful to actually *change* the number. If we want to make the PIC skip two lines we can *add* the number 2 to the program counter:

0043	**movlw**	**d'2'**	**; adds the decimal number 2**
0044	**addwf**	**PCL, f**	**; to the program counter**
0045	**bsf**	**porta, 0**	**; turns on green LED (useless - not**
			; executed)
0046	**bsf**	**porta, 1**	**; turns on yellow LED (useless - not**
			; executed)
0047	**bsf**	**porta, 2**	**; turns on red LED (this one is**
			; executed)

The number in the PCL at address **0044** is **0045**, until the number 2 is added to it. It then becomes **0047**, and so the next instruction to be executed is at address **0047**. The instructions in between are ignored. This particular case is quite useless, as those instructions will *never* be executed, as the number added to the PCL will *always* be 2. However, if the number added to the PCL is *variable*, useful tasks may be performed.

Returning now to a more detailed look at subroutines, you will see the importance of the program counter. When a subroutine is *called*, the contents of the program counter are placed in a special storage system called the *stack*. You can think of the stack as a stack of papers, so when the subroutine is called, the number in the program counter is placed on top of the stack. When a returning instruction is reached, the top number on the stack is placed back in the program counter, thus the PIC returns to execute the instruction after the **call** instruction. In this example we have only used one level on the stack (only one number was placed on the stack). However with the PIC5X series, there are a maximum of *two* levels to the stack (most other PICs have eight). When a subroutine is called within a subroutine, again the number in the PC at the **call** instruction is placed on the top of the stack, pushing the previous number to the level below. The subsequent returning instruction will, as always, select the number on the top of the pile and put it in the PC. If you then call a third subroutine within the second, as you can imagine the third number goes on top, the second is pushed to the bottom level, and the first number is pushed out of the stack (i.e. it is forgotten). This means that it will not be possible to return from the first subroutine, clearly not a desirable situation. The example in Figure 2.16 illustrates this problem.

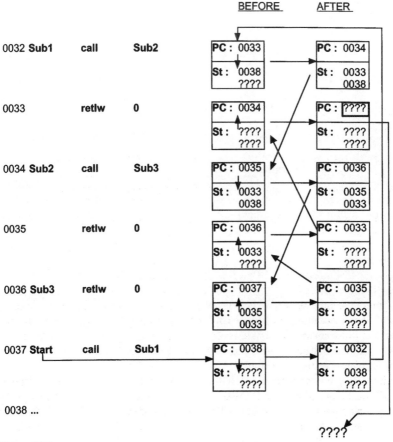

BEFORE   AFTER

0032 **Sub1**	call	Sub2	PC : 0033	PC : 0034
			St : 0038	St : 0033
			????	0038

0033	retlw	0	PC : 0034	PC : ????
			St : ????	St : ????
			????	????

0034 **Sub2**	call	Sub3	PC : 0035	PC : 0036
			St : 0033	St : 0035
			0038	0033

0035	retlw	0	PC : 0036	PC : 0033
			St : 0033	St : ????
			????	????

0036 **Sub3**	retlw	0	PC : 0037	PC : 0035
			St : 0035	St : 0033
			0033	????

0037 **Start**	call	Sub1	PC : 0038	PC : 0032
			St : ????	St : 0038
			????	????

0038 ...

????

Figure 2.16

Begin where it says **Start. W**hen the **call   Sub1** instruction is executed the contents of the program counter are moved into the stack. Then in the subroutine **Sub1**, when the second subroutine (**Sub2**) is called, the contents of the program counter are again moved into the stack, pushing the previous value down a level. Finally upon the third subroutine (**Sub3**) being called, the program counter is again moved into the stack, pushing the second value down a level, and the first *out of the stack*. The PIC has thus forgotten where to return to upon the end of the first subroutine called. Thus at the instruction **retlw   0** in the subroutine **Sub1**, the PIC moves an unknown number **????** from the stack into the program counter, which could make the PIC effectively return anywhere (though this is *probably* the instruction at address **0000**, which is probably the first line of the **Init** subroutine ... but don't count on it!). Do not worry too much about the technical side of all this; what it boils down to is that you can call a subroutine, and you can call a subroutine within a subroutine, but you cannot call a subroutine within a subroutine within a subroutine. Of course, you can call two subroutines within the same subroutine like so:

Sub1	**call**	**Sub2**	;
	call	**Sub3**	;
	retlw	**0**	;
Start	**call**	**Sub1**	;

As you should already know, to return from a subroutine, the instruction is:

> **retlw   number   ;**

This, however, not only returns from a subroutine, but **ret**urns from the subroutine with a literal (**number**) in the working register. This instruction is the key to the encoding process, as the decoding subroutine below demonstrates.

_7SegDisp

	addwf	**PCL, f**	; skips a certain number of
			; instructions
	retlw	b'11101110'	; code for 0
	retlw	b'00101000'	; code for 1
	retlw	b'11011010'	; code for 2
	retlw	b'11111000'	; code for 3
	retlw	b'00111100'	; code for 4
	retlw	b'11110100'	; code for 5
	retlw	b'11110110'	; code for 6
	retlw	b'01101000'	; code for 7
	etc.		
	retlw	b'01010110'	; code for F

Remember, this subroutine is called with the number to be displayed in the working register. This number is then added to the program counter, so the PIC skips that number of instructions. If the number to be displayed is 0, the PIC skips 0 instructions and thus returns from the subroutine with the code to display a 0. This applies for all the numbers being displayed, including the hexadecimal numbers.

Our next project will be a counter. It will count the number of times a push button is pressed, from 0 to F. After 16 counts (when it passes F), the counter should reset. The seven-segment display will be connected to pins RB1 to RB7, and the push button will go to RB0. Figure 2.17 shows the project's circuit diagram; pay particular attention to how the outputs to the seven-segment display are arranged.

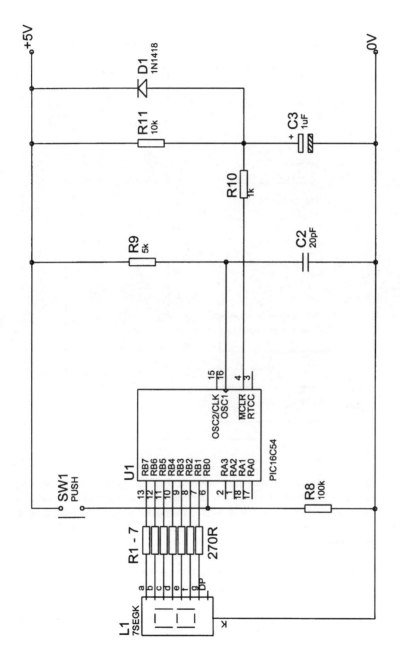

Figure 2.17

The flowchart will be as shown in Figure 2.18.

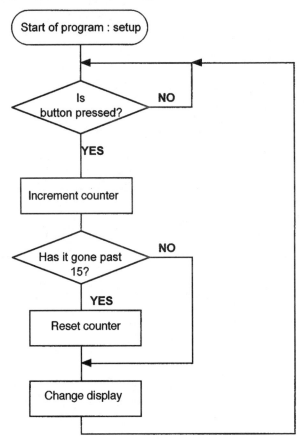

Figure 2.18

The set-up is much like in previous projects, but do not forget to reset any important file registers (such as the one used to hold the number of counts) in the **Init** subroutine. It may also be desirable to move the code for a **0** into Port B at the beginning (rather than simply clearing it). Testing the push button should present no problems either.

Exercise 2.15 What *two* lines will firstly test the push button, and then loop back and test it again if it isn't pressed.

When the push button is pressed the PIC skips out of this loop. In this case the general purpose file register which you are using to keep track of the number of times the button has been pressed (let's call it **Counter**) should be incremented.

Exercise 2.16 What *one* line will accomplish this?

We then need to check to see whether or not more than 15 (F in hexadecimal) counts have been received, or in other words whether or not the number in **Counter** is 16. As you know, the way to see whether or not the number in a file register is a particular value is to subtract that value from the file register (leaving the result in the working register), and then see if the result is zero. This is just like the section we used in the delay program.

Exercise 2.17 **Challenge!** What *four* lines will first test to see whether or not the number in **Counter** has reached 16, and if it has will reset **Counter** to 0 (clear it). Otherwise the PIC should continue, leaving **Counter** unchanged.

Finally we need to change the number in **Counter** into a seven-segment code and move it into Port B, before looping back to **Main**. This is done, as you know, using the encoding subroutine.

Exercise 2.18 Write the *four* lines that should follow the previous four, which take the number from **Counter** into the working register, call the decoding subroutine (name it **_7SegDisp**) which returns with the correct code in the working register, and then move it into Port B. Then the PIC should loop back to **Main**.

Exercise 2.19 Finally, write the subroutine called **_7SegDisp** which contains the correct codes for the seven-segment display.

The program so far is shown as Program F in Chapter 7. It is recommended that you actually build this project. Try it out and you will spot the major flaw in the project.

You should notice that when you press the button, the number 8 will appear on the display, and then when you release the button, the counter will stop on a seemingly random number between 0 and F. This is because the PIC isn't testing for the button to be released. So if you work out roughly how long a cycle takes in the current program when the button is pressed, you can see how often the PIC tests the push button. There are about 11 instructions in the cycle, and the PIC is connected to a 3.82 MHz oscillator. An instruction is executed once every four signals from the oscillator (at 0.96 MHz), so the cycle of 11 instructions is executed at a frequency of about 86 800 Hz, that's 86 800 times a second. So with the current program, if you press the button for one second, counter will count up about 86 800 times (hence the 8 on the display – what you get when the display counts up through all the numbers at high speed). As you can see this project, as it is, would make a good random number generator, but let's move on.

To solve this problem we need to wait until the button is released until we test for it again. The improved program flowchart would be as shown in Figure 2.19.

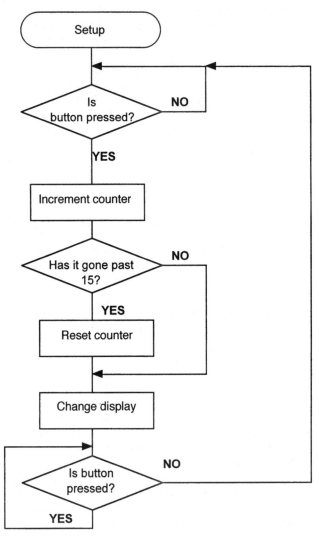

Figure 2.19

All that needs to be changed is that instead of the final line **goto   Main**, we need to test the push button again. The PIC should go back to **Main** if it isn't pressed, and keep looping back if it is pressed.

Exercise 2.20 What *three* lines will achieve this. (**Hint:** You need to give this loop a name).

Assemble the new program (shown again in Program G in Chapter 7), and try it out. Alas, we still have a problem.

You should notice that the counter seems to count up more than once when the push button is pressed (e.g. upon pressing the button it will go from the number 4 to the number 8). This jump varies in size depending on the quality of the push button used. Our problem is due to *button bounce*. The contacts of a push button actually bounce together when the push button is pressed or released. Figure 2.20 shows the signal fed to the PIC.

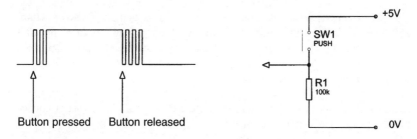

Figure 2.20

The precise details of the bouncing vary according to button type, and indeed may be different every time the button is pressed, but button bounce is always there. As you can see from Figure 2.20 the PIC will count more than one signal, even though the button has only been pressed once. To avoid this, we must wait a short while after the button has been released before we test the button again. This slows down the minimum time possible between counts, but a compromise must be reached.

Example 2.8 To avoid button bounce we could wait 5 seconds after the button has been released before we test it again. This would mean that if we pressed the button 3 seconds after having pressed it before, the signal wouldn't register. This would stop any bounce, but means the minimum time between signals is excessively large.

Example 2.9 Alternatively to attempt to stop button bounce we could wait a hundred thousandth of a second after the button release before testing it again. The button bounce might well last longer than a hundred thousandth of a second so this delay would be ineffective.

A suitable comprise could be about a tenth of a second (as button bounce varies depending on the button you use, this may not be sufficient – so you may have to experiment a little). I am going to choose the longest time possible without having to use more than one postscaler. In this case the oscillator is at 3.82 MHz, divide by four to get 0.96 MHz, and then again by 256 to get the lowest frequency of the TMR0 which equals 3730 Hz. Using my own further postscaler/marker of 255, I can get a frequency of 14.6 Hz. This total time is therefore 0.07 seconds (= 1/14.6) which should be sufficient. The improved program flowchart is shown in Figure 2.21.

The testing instructions are fairly straightforward:

TestLoop   **btfsc**   **porta, 0**   ; tests push button
                 goto   **TestLoop**   ; if pressed keeps looping

However to create a 0.07 second delay, we need to wait for TMR0 to change 255 times. This is not the same as waiting for it to reach 255, because for all we know it could be at 254 anyway. We should therefore reset the TMR0 to zero before the delay. Always be wary of changing the TMR0 because it may be being used by another part of the program.

Exercise 2.21 What *six* lines will create a 0.07 second delay (use a constant marker of 255), and then goto **Main**.

With these new amendments, the project should now work. The final program is shown in Program H in Chapter 7.

Our next project will be a stop clock. It will show minutes (up to nine), tens of seconds, seconds, and tenths of a second, thus requiring four seven-segment displays. Using strobing, these will only require 4 + 7 = 11 outputs. The push button to start, stop and reset the device will require 1 input. In this way the whole project can be squeezed onto the PIC54. RB1 to RB7 will have the seven segment code for *all four* of the displays, RB0 will be the push button, and finally RA0 to RA3 will control the seven-segment displays. The circuit diagram in Figure 2.22 summarizes the setup.

The program flowchart must now be constructed (Figure 2.23).

You should by now be quite familiar with the setup procedure, so it will not be discussed further. Testing for the start button shouldn't present any great difficulties either (use the **btfss** instruction). If the button isn't pressed, the PIC should loop back to **Main**, otherwise it should skip the **goto** instruction and proceed onto the release loop. In this loop the PIC must wait for the button to be released, before skipping on to the main loop. In this main loop the PIC should be constantly updating the time, as well as constantly displaying the four seven-segment displays in turn (the principle behind strobing). On top of all this, it needs to be testing for the *stop* push button, but it mustn't start testing until one second has passed. This is due to button *bounce*, whereby, upon release, the contacts of most buttons bounce, creating the impression to an unsuspecting PIC program that the button is being pressed repeatedly. This would make the

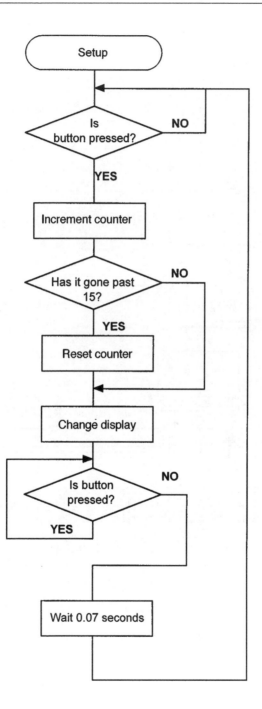

Figure 2.21

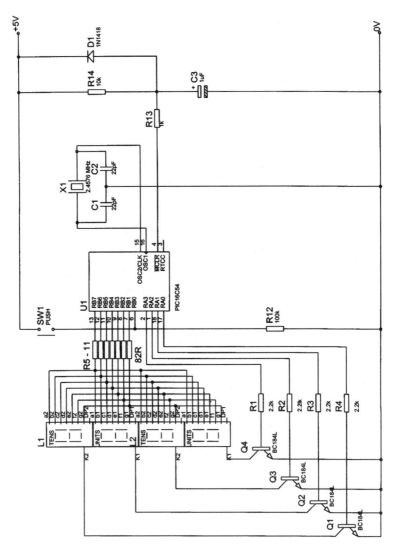

Figure 2.22

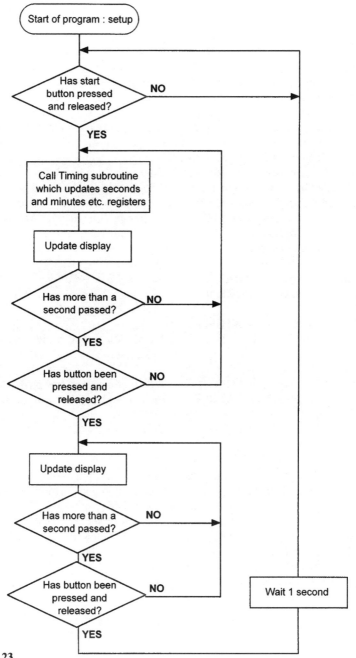

Figure 2.23

stop clock stop almost as soon as it had started, clearly an undesirable property. By waiting one second, we can avoid being plagued by the button bounce, but have to make the sacrifice of making the minimum time that the stop clock can time one second. This shouldn't present too great a problem, as with times of less than one second, human reaction time would have a significant effect.

The timing is best done in a subroutine – this gets it out of the way nicely. However rather than creating a delay, it should merely check whether a certain amount of time has passed, and if it hasn't, continue with other tasks. If the PIC were stuck in a delay subroutine, it would not be able to execute the other parts of the main loop (like the displaying and button testing). This subroutine (let's call it **Timer**), will first have to tell whether or not a tenth of a second has passed, as this is the smallest unit of time being displayed. The TMR0, when prescaled by the maximum amount of 256, counts up 2400 times a second, and thus 240 times in a tenth of a second. Fortunately the number 240 is less than 255, so we can time a tenth of a second using just one marker. We will call this first marker **Mark240**. The first part of the timing subroutine will be reasonably similar to the delay instruction set:

Timer	**movfw**	**Mark240**	; tests to see if TMR0 has passed
	subwf	**TMR0, w**	; through 240 cycles (i.e. 1/10th of a
			; second has passed)
	btfss	**STATUS, Z**	;
	retlw	**0**	; 1/10th of a second hasn't passed, so
			; returns
	movlw	**d'240'**	; 1/10th of a second has passed, so adds
	addwf	**Mark240, f**	; the decimal number 240 to Mark240
	incf	**TenthSec, f**	; increments the number of tenths of a
			; second

Rather than looping back to **Timer**, if the correct time hasn't elapsed the PIC returns from the subroutine, enabling it to go on and perform the other necessary tasks. Also note that the number 240 must have be moved into **Mark240** to begin with (e.g. in the **Init** subroutine). Finally the instruction **retlw** is used to return from a subroutine, but a number must always be specified after it (if the number is of no importance it should be 0 as used above). As is shown above, once a tenth of a second has passed, the file register **TenthSec** is incremented (one is added to it). This is done using the following instruction:

incf	**FileReg, f**	;

This **increments** (adds one to) a file register, leaving the result in the file register. In this way the file register **TenthSec** holds the number of tenths of a second which have passed, and thus will be able to be used easily in the displaying subroutine. (If **TenthSec** counted down from 10 to 0, for example, it *wouldn't* hold the actual number of tenths of a second which had passed.) Once a tenth of a second has passed, we need to check whether a whole second has passed (i.e. if 10 tenths of a second have passed). So we use the technique always used when checking whether a file register has reached a certain number – we subtract that number from the file register, leaving the result in the working register, and then test to see whether or not the result it zero:

```
movlw    d'10'          ; tests to see whether TenthSec has
subwf    TenthSec, w    ;   reached 10 (i.e. whether or not one
                        ;   second has passed)
btfss    STATUS, Z      ;
retlw    0              ; 1 second hasn't passed, so returns

clrf     TenthSec       ; 1 second has passed, so resets
incf     Seconds, f     ;   TenthSec and increments the
                        ;   number of seconds
```

This instruction set is much the same as the one for tenths of a second, except the number we are testing for will always be 10, and we reset back to 0 when the correct time has elapsed. Further sections for tens of seconds and minutes will take much the same form as the one above.

Exercise 2.22 Write the instruction sets to continue the timing subroutine from the line **incf Seconds, f**, for tens of seconds, and then for minutes. (**Hint:** The last line should be **incf Minutes, f).**

The next step is to test to see if **Minutes** has reached 10. At this point the stop clock's maximum is reached, and device should reset – all that is required is clearing **Minutes**, as all the other file register will have reset 'on the way'.

```
movlw    d'10'          ; test to see whether Minutes has
subwf    Minutes, w     ;   reached 10
btfss    STATUS, Z      ;
retlw    0              ; 10 minutes haven't passed, so returns

clrf     Minutes        ; 10 minutes have passed, so resets
retlw    0              ;   Minutes and returns.
```

The entire subroutine **Timer** has been written (in pieces), producing four file registers (**TenthSec**, **Seconds**, **TenSecond** and **Minutes**) which hold numbers to be displayed by the displaying subroutine.

The displaying subroutine has two main tasks: first to choose which display it is going to turn on (tenths of second, seconds, etc.), and second work out what to display on it. As we have a power of two as the number of displays (four is a power of two), we can use a neat trick with the TMR0 to evenly scroll through the different displays. This is the essence of strobing – first one display is turned on for a short period of time with all the others off, then it is turned off and another is turned on with its number displayed. This happens so quickly that we don't even notice it and are given the impression that all are on at the same time. We can use the two least significant bits (bits 0 and 1) of TMR0 to decide which display to turn on. If the two bits in question are **00**, tenths of second are displayed, if they are **01**, seconds are displayed, if they are **10**, tens of seconds are displayed, and finally, if they are **11**, then minutes are displayed. How do we just look at the two least significant bits? How do we ignore the rest of the number? The answer is *ANDing*. The logic command AND takes a certain number of bits as its inputs (in the case of the PIC it takes *two*) and depending on their states creates an output (i.e. the result of the logic operation). The table below (known as a *truth table*) shows the effect of the AND command on two bits.

Inputs		Result
0	**0**	**0**
0	**1**	**0**
1	**0**	**0**
1	**1**	**1**

As you can see, the output bit is high if the first **and** second input bits are also high. A useful property of this command is that if you AND a bit with a **0** the bit is *ignored*, and if you AND a bit with a **1**, the bit is *retained*.

Example 2.10 ANDing the two 8 bit numbers **01100111** and **11110000**, produces the following result:

$$\begin{array}{r} \mathbf{01100111} \\ \underline{\mathbf{11110000}} \\ \mathbf{01100000} \end{array}$$

Notice how by ANDing the top number with **11110000**, bits 4 to 7 are retained (kept the same), whereas bits 3 to 0 have been ignored (replaced with 0). In this way we can ignore bits 2 to 7 of TMR0, retaining only bits 0 and 1.

Exercise 2.23 What number must TMR0 be ANDed with to ignore all but bits 0 and 1?

The instruction that allows us to AND two numbers together is:

andlw   number   ;

This **AND**s the literal (**number**) with the number in the working register. However an alternative instruction more suited to this example is:

andwf   FileReg, f   ;

This **AND**s the number in the working register with the number in a file register, leaving the result in the file register. It would be quite disastrous to actually affect the number in TMR0 as this would mess up the whole of the timing side of things, so we replace the **,f** with a **,w**, so that the result is placed in the working register, leaving the file register unchanged. The instruction pair used to ignore all but bits 0 and 1 of TMR0, leaving the result in the working register is:

movlw   b'00000011'   ; ignores all but bits 0 and 1 of TMR0
andwf   TMR0, w   ; leaving the result in the working
**   ;   register**

How do we use this number to select which display we turn on? We simply add the result (a number between 0 and 3) to the program counter, and have several jumping (**goto**) instructions afterwards which are executed depending on the result:

addwf   PCL   ; adds the result to the program
**   ;   counter**
goto   Display10th   ; displays tenths of a second
goto   Display1   ; displays seconds
goto   Display10   ; displays tens of seconds
goto   DisplayMin   ; displays minutes

The PIC thus branches out to different sections depending on the two least significant bits of the TMR0. These sections will take the following form:

Display10th movfw   TenthSec   ; takes the number out of TenthSec
**   call   _7SegDisp   ; converts the number into 7 seg**
**   ;   code**
**   movvwf portb   ; displays the value through Port B**

**   movlw   b'0010'   ; turns on correct display**
**   movwf   porta   ;**

**   retlw   0   ; returns**

You may have noticed that for a brief time, the *wrong* number is being displayed on a display; this is of no consequence as it is on the wrong display for about a 300 000th of a second. If you are a perfectionist, or find in other cases that there is a considerable delay between putting the correct number in Port B, and turning on the correct display, simply clear Port A after changing the number in Port B. No display is better than a wrong display (for a short period of time). The other sections will be like this, except with a different file register used as the source of the number being displayed, and a different number being moved into Port A.

Exercise 2.24 Write the other three sections required to finish the **Display** subroutine.

The entire **Display** subroutine is now also complete.

The next step is the section where we check for the stop button (only after one second has passed). The easiest method is to write a new time checking subroutine to check for one second. Such a new subroutine could be called **OneSec** and would return if a second hasn't passed, and clear a bit called **sec** when a second does pass:

```
OneSec    movfw    Mark80        ; has 1/30th of a second passed?
          subwf    TMR0, w       ;
          btfss    STATUS, A     ;
          retlw    0             ; no, so returns
          movlw    d'80'         ; yes, so resets postscaler
          addwf    Mark80, f     ;
          decfsz   Post30, f     ; has 1 second passed?
          retlw    0             ; no, so returns

          bcf      sec           ; tells rest of program that 1
                                 ;    second has passed
          movlw    d'30'         ; resets first postscaler
          movwf    Post30        ;
          retlw    0             ; returns
```

The bit **sec** can easily be checked in the main body of the program using the **btfss** instruction – please note that it should be set to start with (this can be done in **Init**). What is the bit called **sec**? As with general purpose register, we can assign general purpose bits. These are bits from general purpose registers. (e.g. **sec** may be bit **0** from file register **08**). It makes little sense to write:

```
          bsf      08, 0         ; etc.
```

Instead, it makes far more sense to the writer to replace **08, 0** with something people will understand, which brings us on to a new command:

#define   name   **FileReg, Bit**

This assigns a particular name to a particular bit in a file register. This doesn't have to be a general purpose file register either – you can rename a bit in one of the ports as the component it is attached to. The fundamental difference between this command and **equ**, is that a *number* must always follow **equ**, whereas *anything* can follow **#define**.

Example 2.11
        #define   LED1   **porta, 0**
etc.
        bsf   LED1   ; turns on first LED (connected to RA0)

However in the case of a general purpose bit, we naturally need to assign it a bit in a general purpose file register (and of course one which we aren't already using, e.g. **Mark80**). I advise having one file register set aside to house all the general purpose bits (you seldom need more than 8), and calling this file register **General** (or a more inspiring name if you can think of one). To define the bit **sec** the following would be written:

        #define   sec   General, 0

If we were to write this, we would naturally have to define the file register **General**:

        General   equ   08

You may then ask (and rightly so) why don't I simply write:

        #define   sec   08, 0

The reason for this is that if I define the file register **General** as number **08**, along with all the other general purpose file registers, there is less danger of actually assigning file register **08** to another specific task by accident. As well as this, people tend to feel more comfortable with names rather than numbers, so it is a good idea to use them were you can. Finally, defining of bits should take place immediately after the defining of file registers in the *declarations* section of the template.

Now that we have a bit which is set when more than one second has passed, testing for the stop button if one second has passed is simply:

        btfsc   sec   ; has 1 second passed?
        goto   MainLoop   ; no, so loops back

```
        btfss      portb, 0        ; has stop button been pressed?
        goto       MainLoop        ; no, so loops back

        etc.                       ; yes so stops counting
```

This concludes the final part of the main loop of the program; all that remains is what happens when the stop button is pressed. The PIC then enters a loop where the display still continues, but where there is no further counting. In this loop we must wait for the button to be released. After the button has been released the PIC should start checking for the reset button, however this checking must not take place until one second has passed. For this to happen we would have to first set **sec**, thus resetting it, and then enter a loop in which we are calling **OneSec**, and then testing for **sec** to be cleared. If cleared, we can skip the loop and keep testing for the reset button:

```
            bsf        sec          ; resets the sec bit
Release2    call       Display      ; displays the finished time
            btfsc      portb, 0     ; waits for button to be released
            goto       Release2     ;

Debounce    call       OneSec       ; has one second passed?
            btfsc      sec          ;
            goto       Debounce     ; no, so loops back

ResetLoop   call       Display
            btfss      portb, 0     ; yes, so tests reset button
            goto       ResetLoop    ; it isn't pressed, so loops back
```

The loop to wait for the reset button is now complete, and all that is left is waiting one second before going back to the start of the program where the start button is tested. We can use the same technique to wait for another second, and then jump back up to **Main**:

```
            bsf        sec          ; resets the sec bit
Release3    call       Display      ;
            btfss      portb, 0     ; is button released?
            goto       Release3     ; no, so loops back

FinalLoop   call       Display
            call       OneSec       ; has one second passed?
            btfsc      sec          ;
            goto       FinalLoop    ; no, so loops back
            goto       Main         ; yes, so loops back to the beginning
```

The entire program (it's quite a large one!) is now complete and is shown in its entirety in Program I in Chapter 7. You will, I hope, find the end result much more satisfying than previous examples, but will recognize a lot more work went into it. When constructing a program of that size (or larger) I cannot stress enough the importance of taking breaks. Even when it is really flowing and you are really getting into your program, if you step back for ten minutes and relax, you will return looking at the big picture, and may find you are overlooking something simple. Good planning with flowcharts and diagrams will help prevent such over-sights significantly. You should also talk to people about decisions you should make along the way – even if they may not know the answer any more than you do, simply asking the question and talking it through helps you get it straight, and the majority of the time you will end up answering your own question.

Logic gates

After a long and complicated project, let's return to something simpler. You've now seen three logic gates (inclusive OR, exclusive OR, and AND), and we'll now look at the other five (NOR, NAND, BUFFER, NOT and XNOR). The truth tables for the new gates are as follows:

NOR

Inputs		Result
0	0	1
0	1	0
1	0	0
1	1	0

The result is the opposite of an inclusive OR gate (i.e. **not** an inclusive OR gate).

NAND

Inputs		Result
0	0	1
0	1	1
1	0	1
1	1	0

The result is the opposite of an AND gate (i.e. **not** an AND gate).

BUFFER

Input	Result
0	0
1	1

Only one input is used, the output copies the input.

NOT

Input	Result
0	1
1	0

Again only one input, but the output is the opposite of the input (i.e. **not** the input).

XNOR

Inputs		Result
0	0	1
0	1	0
1	0	0
1	1	1

The result is the opposite of an exclusive OR gate (i.e. **not** an exclusive OR gate).

There aren't instructions for all these gates, but all can be constructed through a combination of those which we are given. The project to experiment with the use of these gates and their instructions will be a multi-gate IC (a chip which will effectively act as any of these eight gates). There will be two inputs and one output which are the actual parts of the artificial gate. There will also be three bits for choosing which gate. Three bits can select a total of eight variations (000 to 111). There will be one combination for each of the eight logic gates. These selection bits will be RA1 to RA3, and the inputs of the gate will be RB0 (main input) and RA0 (secondary input), with the gate output at RB4. The circuit diagram is shown in Figure 2.24.

The flowchart must now be constructed.

Exercise 2.25 Have a go yourself at constructing the flowchart, before looking at my version in the answer section (Appendix H). Remember, as long as the gist of it is the same, it isn't crucial that the minor details are the same as mine, but you need not make it more than three boxes in size, as we aren't yet concerned with sorting out how to manage the imitating of the individual gate types.

First the PIC must decide which logic gate it is meant to imitate. This can be done using the branching technique of adding a number to the program counter, which you are already familiar with. Unfortunately the number in Port A is not the number we want to add to the program counter (remember one of the logic gate inputs is in Port A). We could simply ignore bit 0 (using ANDing), but this would leave us with a number which could go up to 14 (0000 to 1110). What we really want to do is *rotate* the number to the right (making 1110 into 0111). Fortunately there exists such a rotating instruction:

rrf   FileReg, f   ;

This rotates the number in a file register to the right, leaving the result in the file register. Its complementary instruction is:

rlf   FileReg, f   ;

This rotates the number in a file register to the left, again leaving the result in the file register. You may wonder where the bit that gets 'bumped off' goes, and

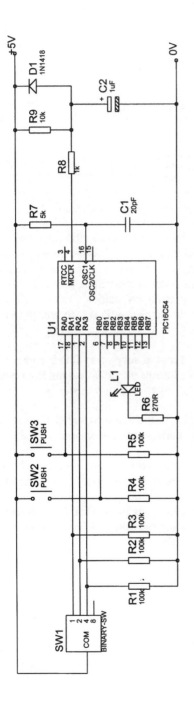

Figure 2.24

where the bit that fills the gap left comes from. There is an intermediate bit called the *carry flag*. This is one of the flags (like the zero flag) in the STATUS register. It has other purposes as well as that shown in Figure 2.25.

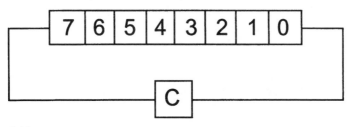

Figure 2.25

So when rotating right, the state of bit 0 is moved into the carry flag, and the previous state of the carry flag is moved into bit 7. This is a consequence of the carry flag's main property which will be discussed at a later stage. It is important to clear the carry flag before any rotation instruction, because otherwise, if set, it will put a one where a gap was left upon rotation – in most cases this in undesirable. To thus be able to use the number in Port A to branch to the correct place, the following is done to it:

Main bcf STATUS, C ; makes sure carry flag is clear
rrf porta, w ; bumps off bit 0, leaving the result in
; the working register

The number in the working register is then simply added to the program counter, and the PIC will skip to the correct **goto** instruction:

movfw porta ; takes the number out of Port A
addwf PCL, f ; adds this number to the program
; counter
goto NOT ; the code for a NOT gate is received
goto BUFFER ; the code for a BUFFER is received
goto AND ; etc.
goto IOR ;
goto XOR ;
goto NAND ;
goto NOR ;
goto XNOR ;

Now each of these sections must be examined separately. For the NOT gate, only one of the inputs is used (the one at RB0); this is simply toggled. As you should know this can be done using the XOR logic operation. To toggle bit 0, simply exclusive OR Port B with **00000001**. This result must be stored in a file register (not Port B). In this case selective toggling is not necessary and the entire file register may be inverted using the **comf** instruction. If bit 0 of the file register is high, then the output bit (bit 4) should be set and vice versa. This could be done using testing instructions (**btfss**) and setting/clearing instructions (**bsf** and **bcf**), but a more cunning method employs the following command:

>   **swapf   FileReg, w   ;**

This **swap**s the lower nibble (bits 0 to 3) with the upper nibble (bits 4 to 7) of a file register and leaves the result in the working register.

Example 2.12

>   **movlw   b'00110101'   ; moves a number into file register ABC**
>   **movwf   ABC   ;**
>   **swapf   ABC, w   ; swaps the nibbles of ABC**

The number in the working register at the end of the three lines would be **01010011**.

Exercise 2.26 What would be the resulting number if the following number were 'swapped': **00000001**?

Thus if we swap the file register holding the result of the toggle operation, the state of bit 0 will effectively be swapped with that of bit 4 (as Exercise 2.26 has just shown). The result of this swapping (left in the working register) would then be moved into Port B:

>   **comf   portb, w   ; inverts Port B leaving the result in the**
>   **; w. reg.**
>   **movwf   STORAGE   ; stores the result in a file reg. called**
>   **; STORAGE**
>   **swapf   STORAGE, w   ; swaps the nibbles of STORAGE**
>   **movwf   portb   ; sends the result back to portb**

A logical process has been followed and the operation has been executed in four lines. However by changing the process at which we arrived at the result, a shorter and less wasteful method can be used. Instead of inverting the register and *then* swapping the nibbles, which requires the use of a storage file register, we could first swap the nibbles of Port B, and invert it. This would require no intermediate file register:

```
        swapf    portb, f      ; swaps bit 0 for bit 4 (input for output)
        comf     portb, f      ; inverts the whole of port b (and thus
                               ;   also bit 4)
        goto     Main          ; loops back to Main
```

The result is half the size and requires no extra file registers. The buffer gate will be even easier as it is simply a NOT gate with no inverting:

```
        swapf    portb, f      ; swaps the input bit for the output bit
        goto     Main          ; loops back
```

An AND gate will simply involve ANDing Port A with Port B and the resulting bit 0 will be the output. Thus the result should be stored, swapped and then moved back into Port B. There is no shortcut possible this time, as Port A possesses no bit 4.

```
        movfw    porta         ; takes the number out of Port A
        andwf    portb, w      ; ANDs the number with Port B
        movwf    STORE         ; stores result in a GP file register
        swapf    STORE, w      ; swaps nibbles and put result in
                               ;   working reg.
        movwf    portb         ; returns result to Port B
        goto     Main          ;
```

An inclusive OR is basically the same:

```
        movfw    porta         ; takes the number out of Port A
        iorwf    portb, w      ; inclusive ORs the number with Port B
        movwf    STORE         ; stores result in a GP file register
        swapf    STORE, w      ; swaps nibbles and put result in
                               ;   working reg.
        movwf    portb         ; returns result to Port B
        goto     Main          ;
```

Both these have the same last four lines, and thus these could be separated from the two, and then jumped to when the correct place is reached. This would involve adding a **goto** instruction after the **iorwf** instruction jumping to the line after the **andwf** instruction, and would save three lines. The AND section would therefore be:

```
AND     movfw    porta         ; takes the number out of Port A
        andwf    portb, w      ; ANDs the number with Port B
```

common	movwf	STORE	; stores result in a GP file register
	swapf	STORE, w	; swaps nibbles and put result in
			; working reg.
	movwf	portb	; returns result to Port B
	goto	Main	;

And the OR section simply would be:

	movfw	porta	; takes the number out of Port A
	iorwf	portb, w	; inclusive ORs the number with Port B
	goto	common	; jumps to shared section in the AND
			; section

An exclusive OR could be:

	movfw	porta	; takes the number out of Port A
	xorwf	portb, w	; exclusive ORs the number with Port B
	movwf	STORE	; stores result in a GP file register
	swapf	STORE, w	; swaps nibbles and put result in
			; working reg.
	movwf	portb	; returns result to Port B
	goto	Main	;

Remember, if one bit is set, the other is toggled, and if it is clear, the other is kept the same (so it is a NOT or a BUFFER gate). We could use this property to try to slim down this section. First begin with the line which is common to both gates, and then test the other input bit so see whether or not to invert the output:

	swapf	portb, f	; moves bit 0 into bit 4
	btfsc	porta, 0	; tests other input bit
	comf	portb	; if it is set, the gate should invert,
			; otherwise it doesn't
	goto	Main	;

In this way the five lines have been cut down to three, and a general purpose file register didn't need to be used. Convinced? You shouldn't be. Remember that because some of the lines in the AND and IOR were repeated, they were only put down once and then accessed by both. The same could be done with the XOR section were we to use the longer method. In this way the XOR section will actually be shorter if the seemingly long method is employed (work it out for yourself!). Although on its own, the previous method is the best for XORing, the following section will be shortest in this program:

```
        movfw    porta        ; takes the number out of Port A
        xorwf    portb, w     ; exclusive ORs the number with Port B
        goto     common       ; jumps to shared section in the AND
                              ;   section
```

A NAND gate is exactly like an AND except that the result is NOTed just before it is moved into Port B. We are going to perform the same doubling up trick as before, and so have labelled the line after **andwf**, **common2** (we can't label two things with the same name, because otherwise the PIC wouldn't know which one to go to):

```
            movfw    porta        ; takes the number out of Port A
            andwf    portb, w     ; ANDs the number with Port B
common2     movwf    STORE        ; stores result in a GP file register
            swapf    STORE, f     ; swaps nibbles
            comf     STORE, w     ; inverts result and puts result in
                                  ;   working reg.
            movwf    portb        ; returns result to Port B
            goto     Main         ;
```

The NOR (NOT inclusive OR) is similar:

```
        movfw    porta        ; takes the number out of Port A
        iorwf    portb, w     ; inclusive ORs the number with Port B
        goto     common2      ;
```

Again these two have the same last four lines, so the same can be done as was done with the AND and OR sections. The XNOR section is like the exclusive OR section, just two lines and then a jump to the line after **andwf** in the NAND section.

```
        movfw    porta        ; takes the number out of Port A
        xorwf    portb, w     ; exclusive ORs the number with Port B
        goto     common2      ; jumps to the middle of the NAND
                              ;   section
```

The program is now complete and the whole lot is shown in Program J in Chapter 7.

The watchdog timer

One of the useful properties of the PIC is its *watchdog timer* – an on board timer which is driven by an resistor/capacitor network which is actually in the PIC. It is thus completely independent to outside components. The watchdog timer steadily counts up, and when it reaches its maximum, the PIC will automatically reset. It is thus quite useful in devices where it is not a great problem to be constantly resetting (for at least most of the time), e.g. alarm systems. The time for the watchdog timer to cause a *timeout* (for it to make the PIC reset) varies from 18 milliseconds to 2.3 seconds depending on the amount of prescaling. You can prescale it using the option register (you may remember this from when we studied the TMR0). If left unprescaled it will cause a timeout after 18 ms. To prescale it, bit 3 of the option register must be set, thereupon bits 2 to 0 decide how much it is prescaled by (Table 2.2).

Table 2.2

PS2	PS1	PS0	Prescaling rate
0	0	0	1:1
0	0	1	1:2
0	1	0	1:4
0	1	1	1:8
1	0	0	1:16
1	0	1	1:32
1	1	0	1:64
1	1	1	1:128

The maximum prescaling (128) will cause it to timeout after (0.018×128) seconds = 2.304 seconds. There is, however, no way to simply turn the watchdog timer off unless you don't need it at any period in the program, in which case it is turned off as the PIC is blown. If it is needed for part of the program, how do you stop it from affecting the PIC for the rest of the program? The answer is constantly resetting it. The instruction for this is:

clrwdt   ;

This **clears** the **watchdog timer** (i.e. makes it 0), and thus resets it. This must be done at specific intervals to stop the watchdog timer reaching its maximum and thus causing the timeout, i.e. if the watchdog timer resets the PIC after 18 ms, then you need the **clrwdt** instruction to be executed at least once every 18 ms.

To try out the watchdog timer, the next project will be an alarm system. There will be a signal coming from a motion sensor at RA0 (it can be simulated by a push button), and a siren (or buzzer) at RA3 to indicate when the alarm has been set off. A toggle switch (RA1) will either set, or disable the alarm, a green LED (RB0) will show the alarm is disabled, and a red LED (RB1) will show it to be set. To conserve battery life the LEDs will flash rather than stay turned on, flashing on for one tenth of a second every 2.3 seconds (this number should sound familiar). Once triggered, the siren will go on indefinitely until the device is turned off. You may want to make an addition whereby it turns off after 20 minutes, but this is not investigated in this example. The circuit diagram is shown in Figure 2.26.

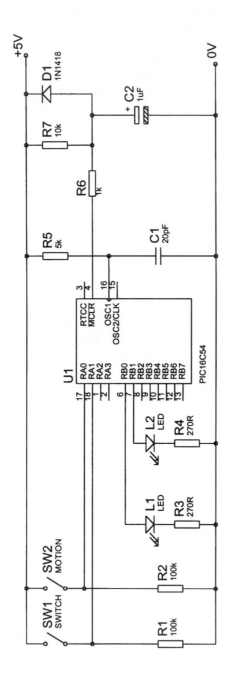

Figure 2.26

The flowchart for the program is as shown in Figure 2.27.

As in the previous example, you are now expected to write most of the program, but naturally you will be guided through each step.

Exercise 2.27 Write the *three* lines used to first test the setting switch, and thereupon jump to another part of the program called **GreenLed** if the bit is high or simply turn on the red LED if the bit is low (as shown in Figure 2.27).

Exercise 2.28 Write the *two* lines which make up the section called **GreenLed**, in which the green LED is turned on, and then the PIC jumps back to a section labelled **TenthSecond**.

Exercise 2.29 Write the *seven* lines which will make up the section which tests to see whether or not a tenth of a second has passed, and turns off all LEDs if such a time has passed. In either case it then moves on to the rest of the program. Don't forget that as you are using the prescaler for the WDT, TMR0 is not prescaled. You will therefore need to slow it down by 256 (a task which is normally done by the prescaler) yourself. This is best done by moving TMR0 into the working register, and then testing the zero flag. If it is set, the TMR0 has reached zero, and you may continue on to the next postscaler (after incrementing TMR0), otherwise skip everything by going to the next section labelled **Continue**. This next postscaler should be around 240 because 2400/10 = 240, but could vary depending on what is easiest. Do not forget to reset the postscaler after it has reached 0; however the correct number should be moved into it to start with in the **Init** subroutine. Having said this, if your postscaler is 256, you don't need to reset it ... think about it.

The next step is to test the setting button once more, to see whether or not it should react to the alarm being triggered. If the alarm is disabled (the bit is high), the PIC should return to the **TenthSecond** section, otherwise it should continue.

Exercise 2.30 What *two* lines will achieve this?

If the PIC continues, the alarm is set, and the trigger bit (RA0) should be tested. If no signal is received, the PIC should loop back to **TenthSecond**, or otherwise continue.

Exercise 2.31 What *two* lines will achieve this?

If the motion sensor has been set off the siren should be turned on, and the PIC should enter a cycle where the watchdog timer is constantly being reset.

Exercise 2.32 What *three* lines will finish the program?

As always the whole program is shown in Program K.

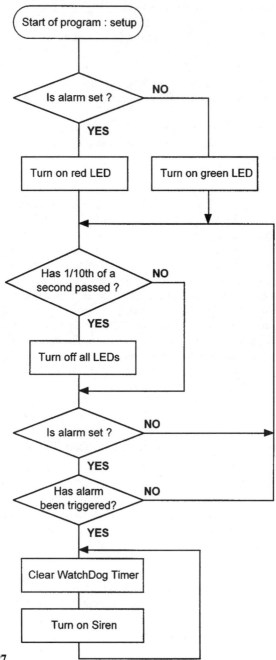

Figure 2.27

Final instructions

There are only four more instructions which you haven't yet come across. You should be able to guess the functions of the first two of these – **decf** and **incfsz** – as they are just like their counterparts.

decf   **FileReg, f**   ;

This **dec**rements (subtracts one from) the number in a file register, leaving the result in the file register.

incfsz   **FileReg, f**   ;

This **inc**rements (adds one to) the number in a file register leaving the result in the file register. If this result is zero the PIC will skip the next instruction.

The next instruction may seem absolutely pointless but *does* actually come in quite handy every now and then:

nop   ;

This stands for **no op**eration, and does nothing.

Finally, if you are tired by now, you'll be pleased to learn the last instruction to be learnt is:

sleep   ;

As you may have guessed, this sends the PIC to sleep (a special low power mode). The outputs will stay the same when the PIC goes into sleep, and can be woken up by a watchdog timer timeout, or an external reset (from the $\overline{\text{MCLR}}$ pin). A useful application combining both the **sleep** instruction and the watchdog timer allows devices to appear to automatically turn on. If, for example, a device were to turn on when moved, the PIC should test a vibration switch, go to sleep (until reset by the watchdog timer) if there is no movement, or alternatively skip out of the loop and constantly reset the watchdog timer as it carries on through the rest of the program, if there is movement. In this way the PIC would be in a low power consuming mode for most of the time (it is effectively off), and would come to life when movement is detected. A Figure 2.28 demonstrates this best.

Indirect addressing

All the instructions have now been studied, but there remains one more concept – that of *indirect addressing*. You may have noticed two file registers (*indirect address* (**00**) and *FSR* (**04**)) have not been explained yet, and these are both involved in this concept. This is probably the hardest idea to fully grasp and so it will be explained twice. First I will introduce it technically, then give an analogy which should make it easier to understand.

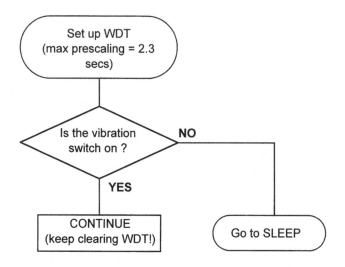

Figure 2.28

Think of storing a number (**N**) in a general purpose file register; you would move the number **N** into (for example) file register number **09**. This is *direct addressing*. However you could also tell the PIC to move the number **N** into file register number **X**, where the file register called **X** holds the number **09**. This is called *indirect addressing*. The file register **X** is actually called the **file select register** (because it is a file **register** which **selects** which **file** register to move a number into). To use indirect addressing, move the number you wish to be stored into the **indirect address**. The indirect address is therefore not a file register as such, merely a gateway to another file register.

If you are still confused by this stage (I don't blame you), the following analogy should set things straight. Think of the **indirect address** as a envelope, and the **file select register** as the address on the envelope. When you use indirect addressing you put the number in an envelope, and it is sent to the address on the envelope (just as with our own reliable post service except with a delivery time of roughly 0.000001 seconds it is slightly faster!).

Example 2.13 Move the number **00** into file registers numbers **08** to **1F**.

Rather than writing:

clrf	**08**	**; clears file register number 08 (it hasn't**
		; been given a name)
clrf	**09**	**; clears file register number 09**
clrf	**0A**	**; clears file register number 0A**
etc....		
clrf	**1F**	**; clears file register number 1F**

... we can use indirect addressing to complete the job in fewer lines. First the address we want to affect is **08**, so we should move **08** into the **file select register** (the address on the envelope):

movlw	**d'08'**	**; moves the number 08 into the FSR**
movwf	**FSR**	**;**

We then send the number **00** through the 'post' by moving it into the **indirect address** (the envelope). The instruction **clrf** effectively moves the number **00** into the file register (thus clearing it):

clrf	**INDF**	**; clears the indirect address**

File register **08** has now been cleared (whatever you now do to the **INDF** you actually do to file register number **08**). We now want to clear register **09**, we thus increment the **FSR** (add one to it), so now whatever you do to the **INDF** you actually do to file register number **09**.

incf	**FSR**	**; increments the FSR**

The PIC can now loop back to the line where the **INDF** is cleared. However it must first check to see whether or not the FSR has passed the file register **1F**, in which case it should jump out of the loop. To see whether a file register holds a particular number, you subtract that number from the file register and see whether or not the result is zero:

movlw	**20h**	**; has the FSR reached the hexadecimal**
subwf	**FSR, w**	**; number 20?**
btfss	**STATUS, Z**	**;**
goto	**ClearLoop**	**; it hasn't, so keep looping**
		; it has, so exits loop

The following instruction set is very useful to put in the **Init** subroutine to systematically clear a large number of file registers:

```
            movlw    d'08'         ; moves the number 08 into the FSR
            movwf    FSR           ;
ClearLoop   clrf     INDF          ; clears the indirect address
            incf     FSR           ; increments the FSR
            movlw    20h           ; has the FSR reached the
            subwf    FSR, w        ;   hexadecimal number 20?
            btfss    STATUS, Z     ;
            goto     ClearLoop     ; it hasn't, so keep looping
                                   ; it has, so exits loop
```

You can adjust the starting and finishing file registers (at the moment **08** and **1F** respectively). If you want to clear a complete block of file registers *starting from file register number 05* (Port A), a shorter (and quite cunning) instruction set may be used. Rather than starting from the lowest number file register, start from the highest (e.g. **1F**), and then keep decrementing the **FSR**. Here comes the complicated part – when the **FSR** is decremented to **04** (the address of the **FSR**), whatever happens to the **INDF** happens to the **FSR**. When we then clear the **INDF**, we actually clear the **FSR**. We can test for this using the zero flag. In the loop, rather than testing to see whether or not the **FSR** has reached a certain number, we simply move it into the working register, and then test the zero flag. Because the instruction segment decrements the FSR *before* clearing the INDF for the first time, we need to move a number one greater than the first file register we wish to clear into the FSR to begin with:

```
            movlw    20            ; moves the hex number 20 into the
            movwf    FSR           ; FSR
ClearLoop   decf     FSR           ; decrements the FSR
            clrf     INDF          ; clears the indirect address
            movfw    FSR           ;
            btfsc    STATUS, Z     ; is the FSR zero?
            goto     ClearLoop     ; no, so keeps looping
                                   ; yes, so exits looping
```

This instruction set therefore does *not* clear file register **03** (STATUS), **02** (PCL), **01** (TMR0), and **00** (INDF). This is desirable, particularly as clearing the PCL would make the next instruction to be executed the one at address **00** (the first line after the instruction **org   0** – probably the first line of the **Init** subroutine).

The STATUS file register

Just before we move on to the final program in this chapter, we will examine the STATUS file register in greater detail.

The STATUS file register

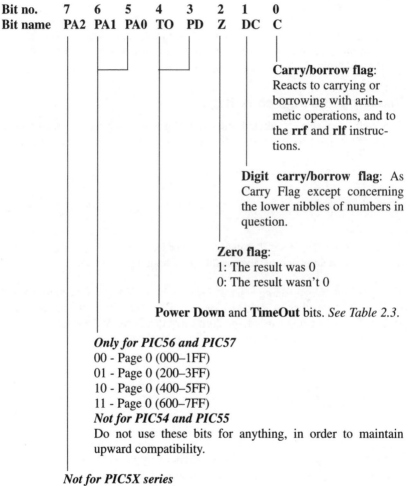

Bit no.	7	6	5	4	3	2	1	0
Bit name	PA2	PA1	PA0	TO	PD	Z	DC	C

Carry/borrow flag: Reacts to carrying or borrowing with arithmetic operations, and to the **rrf** and **rlf** instructions.

Digit carry/borrow flag: As Carry Flag except concerning the lower nibbles of numbers in question.

Zero flag:
1: The result was 0
0: The result wasn't 0

Power Down and **TimeOut** bits. *See Table 2.3.*

Only for PIC56 and PIC57
00 - Page 0 (000–1FF)
01 - Page 0 (200–3FF)
10 - Page 0 (400–5FF)
11 - Page 0 (600–7FF)
Not for PIC54 and PIC55
Do not use these bits for anything, in order to maintain upward compatibility.

Not for PIC5X series
Do not use these bits for anything, in order to maintain upward compatibility.

Table 2.3 Power Down and TimeOut bits

TO	PD	Reset caused by...
0	0	WDT wakeup from sleep
0	1	WDT timeout (not during sleep)
1	0	MCLR wakeup from sleep
1	1	Power - up

There are three new concepts introduced: the *digit carry flag*, the business of *pages* of memory, and the two bits which we can test to find the reason behind the PIC resetting.

The carry and digit carry flags

The digit carry flag is affected only by addition and subtraction instructions. Think of the numbers in question (being added or subtracted) in hexadecimal.

```
    C   DC
        X X
  +     X X
        X X
```

The digit carry flag is set if something is carried over when adding the lower nibbles of two numbers together, and clear if nothing is carried.

Example 2.14 When adding 56h and 3Ah, we first add the lower nibbles: A and 6. These add together to make 16, or in other words, leave 0 and carry a 1. Because a one *is* being carried, the digit carry flag is *set*. We now add 5, 3, and 1 (carried over) making 9. *Nothing* is carried over so the carry flag remains *low*.

```
    0   1
        5 6
  +     3 A
        9 0
```

Example 2.15 When adding 32h and F5h, we first add the lower nibbles: 2 and 5. These add together to make 7, or in others words, leave 7 and carry nothing. Because *nothing* is being carried, the digit carry flag is *clear*. We now add 3 and F making 18, or in other words 2 and carry a 1. Because a one *is* being carried, the carry flag is *set*.

```
    1   0
        3  2
+       F  5
        2  7
```

When subtracting, both act as $\overline{\text{borrow}}$ bits, i.e. if something *is* borrowed when subtracting, they are *clear* and vice versa. (The bar over the name, as with the $\overline{\text{MCLR}}$, means that it is active low – triggered by a negative result). The digit carry ($\overline{\text{borrow}}$) flag again concerns the lower nibbles, and the carry ($\overline{\text{borrow}}$) flag the upper nibbles.

```
      cX      dcX
-      X        X
       X        X
```

Example 2.16   When subtracting 6Bh from 8Dh, we first subtract the lower nibbles (B from D). These leave 2, borrowing nothing. Because *nothing* is borrowed, the digit carry ($\overline{\text{borrow}}$) flag is *set*. We now subtract 6 from 8, leaving 2 and borrowing nothing. The carry ($\overline{\text{borrow}}$) flag is therefore also set.

```
     08  0D
-     6   B
      2   2
```

Example 26:   When subtracting 7Eh from 42h, we first subtract the lower nibbles (E from 2). We need to borrow 1, making the subtraction $\overline{12h} - E$. This leaves 4, borrowing 1. Because one *is* borrowed, the digit carry ($\overline{\text{borrow}}$) flag is *clear*. We now subtract 7 from 3 (4 – 1 which was borrowed). We again need to borrow 1, making the subtraction $\overline{13h} - 7$. This leaves C, borrowing 1. Because one *is* borrowed, the carry ($\overline{\text{borrow}}$) flag is therefore also *clear*.

```
     1(4–1)    12
-      7        E
       C        4
```

This result is effectively a negative number (C4 in this case corresponds to –3C).

To summarize the effect of subtraction on the carry flag: if the result is negative it is clear, and if it is positive (or zero) it is set. The same applies to the digit carry flag except that you look at the lower nibbles, rather than the whole number when performing the subtraction.

Pages

We turn now to this business of pages. You may remember from studying the program template that the PIC54 and PIC55 have 1FF bytes of memory (they can have 1FF instructions). Other members of the PIC5x series can have more than this: the PIC56 has 3FF, and the PIC57 can have up to 7FF. From this we can see that the program counter in the PIC54 and PIC55 is 9 bits long (111111111 = 1FFh), 10 bits long in the PIC56 (1111111111 = 3FFh), and 11 bits long in the PIC57 (11111111111 = 7FFh). The memory of the program (which holds the instructions) is organized as shown in Figure 2.29.

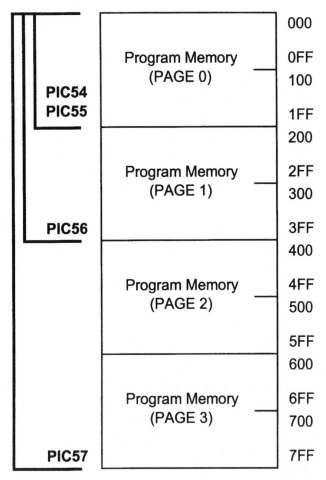

Figure 2.29

We will look first at these bits in relation to the PIC54 and PIC55. None of the three most significant bits of STATUS concern us, but the bit 8 (the ninth bit) of the program counter is quite significant. For the PIC5x series, the *stack* (used to store the number upon a **call** instruction) is only *8 bits wide*! As well as this, the PIC has only 8 bits to move into the program counter when performing the **call** instruction, so your subroutines must start in the lower half of the program memory (000–0FF).

Example 2.18 The program memory has been allocated in the following way:

01FF		**goto**	**Start**	**; begins at start**
0000	**Init**	**clrf**	**porta**	**; resets I/O ports**
0001		**clrf**	**portb**	**;**
etc.		[ALL SUBROUTINES]		
	Timing			
0101		**decfsz**	**Post20**	**;**
0102		**goto**	**Timing**	**;**
0103		**decfsz**	**Post30**	**; has this happened 600**
				; times?
0104		**goto**	**Timing**	**; loops back**
0105		**retlw**	**0**	**; returns**
0106	**Start**	etc.		
0140		**goto**	**Main**	**; loops back**

END

The program in Example 2.18 will not work because the start of the **Timing** subroutine is not in the lower half of the page (000–0FF). Even though it is the top half of the page as you write it, we call 000–0FF the lower half because the numbers are lower than 100–1FF. 000–0FF can be expressed using only eight bits, but once you pass 100 you need a ninth bit to specify which half of the page you are talking about. This ninth bit is not available when dealing with subroutines, so they need to stay in the lower half. This shouldn't often cause much trouble unless you are dealing with quite a long program. If you tried to call a subroutine which started at address (0101), as in Example 2.18, the PIC would in fact jump to address 0001 instead, and keeping going until it met a return instruction (i.e. the program would *crash*).

The PIC56 is blessed with an extra page of memory (you may want to glance back at Figure 2.29 to refresh your memory). The same business about bit 8 of the program counter, and you must always keep the subroutines in the top half of the page 0. The bit called PA0 (bit 5) in the STATUS register may be read to see which page the PIC is in, and is written to when jumping to another page.

The PIC57 has an impressive four pages and all the rules concerning the PIC56 apply. This time both PA0 and PA1 (bits 5 and 6) of STATUS may be read to see which page the PIC is in, and is again written to when jumping to another page.

There is another problem we face when adding to the program counter over half page boundaries. If the number in the program counter is 0FDh (253 in decimal), and you add 5 to it. The result will be 003h. The result of this is that the PIC would then execute the instruction at address 003, rather than the desired one at address 103. It is therefore necessary to have any such additions entirely in one half page, and not across any boundaries.

What caused the PIC to reset?

The Power Down and TimeOut bits can be read at the beginning of the program to see what made the PIC reset (i.e. why is it at the start of the program). This could simply be due to the fact that it had just been turned on (power-up), or alternatively due to WDT timeout. This may be important because you may not want the PIC to do the same thing (e.g. setting up, or perhaps clearing, of file registers) when it first starts up, as when it is reset by the WDT for example.

Table 2.4

TO(4)	PD(3)	Reset caused by...
0	0	WDT wakeup from sleep
0	1	WDT timeout (not during sleep)
1	0	MCLR wakeup from sleep
1	1	Power-up

Example 2.19 To make the PIC call the **Init** subroutine when first powered up, but not when reset for any other reason (i.e. just skip the **call   Init** line), the following instruction set is used:

Start	**btfsc**	**STATUS, 3**	**; tests PowerDown bit**
	btfss	**STATUS, 4**	**; PD is 1, test TimeOut bit**
	goto	**Main**	**; PD is 0, or TO is 0, so skips Init**
	call	**Init**	**; PD and TO are 1, so calls Init**
Main	etc.		

Exercise 2.33 Make the PIC test to see whether there was a WDT timeout, or see if it's just powered up. If it has just powered up call a subroutine called PreInit, otherwise carry on.

Some useful (but not vital) tricks

1. If you are growing tired of the lengthy **goto** instruction, you may be pleased to read that it can be abbreviated to **b**. The **b** instruction (it stands for **b**ranch) does exactly the same thing as **goto**.

Example 2.20

b	**Start**	; goes to Start

2. Another useful trick enables you to go to a specific part of the program, and then skip any number of instructions. This is done by adding **+1**, for example after the label, in a **goto** instruction.

Example 2.21

	goto	**Start+1**	; goes to Start and skips the next instruction

Start	**call**	**Init**	; sets things up
	bsf	**porta, 0**	; turns on an LED

One warning with this instruction is not to use it too frequently, and avoid large skips (e.g. **+14**). In such cases it is probably a good idea to simply add another label at the place you want to go to. Be wary of going back to your program and adding lines (corrections or afterthoughts, etc.), because the number of lines you need to skip may change.

Example 2.22

	goto	**Start+1**	; goes to Start and skips the next instruction

Start	**call**	**Init**	; sets things up
	bsf	**portb, 0**	; turns on buzzer
	bsf	**porta, 0**	; turns on an LED

If we still want the PIC to skip to the line where the LED is turned on, we will need to remember to change the **+1** to **+2**. Such changes are easy to forget if your program is riddled with long skipping gotos.

Final PIC5x program – 'bike buddy'

Our final program in this chapter on the PIC54 and 55 will tie together many of the ideas covered. The next chapter will deal with the PIC71 which demonstrates the use of *interrupts* and *analogue to digital conversion*. It will be a mileometer and speedometer for bicycles. The device should consist of three seven-segment displays (up to 999 kilometres recorded, and an accuracy of 0.1 kph), a toggle switch to change mode (mileage/speed), and an input from a reed

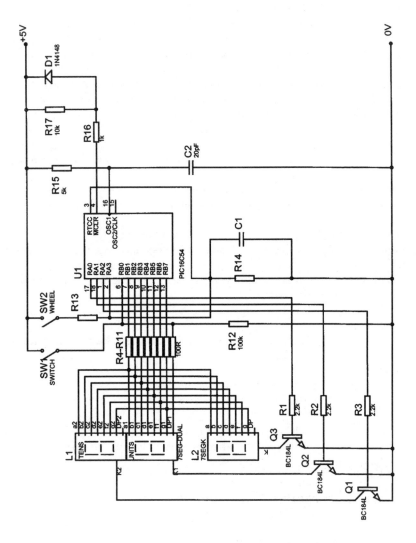

Figure 2.30

switch activated by a magnet on the wheel. This is how speed and mileage are detected. With strobing this makes a total of seven outputs for the seven-segment code (RB1 to RB7), three outputs to select the correct seven segment-display (RA0 to RA2), one input for the toggle switch (RB0), and the input for the reed switch (RA3). This makes a total of 12 I/O, which conveniently just fits on the PIC54. This leaves us with one problem. When displaying the speed, one of the decimal points of the seven-segment display will need to be on, but as we've just worked out there are no spare outputs. However, the decimal point will only need to be on when the toggle switch is selecting the speedometer mode. We can therefore directly link the toggle switch to the decimal point as shown in the circuit diagram in Figure 2.30. The flowchart is shown in Figure 2.31.

I have made this flowchart slightly more detailed than usual because this time you are expected to write the program yourself. If you break things up into the boxes described in the flowchart you should be able to manage everything with little difficulty. If you get stuck, the program I wrote is in Program L in Chapter 7, but remember that the way I did something in my program may not necessarily fit in with your program as the two are likely to have some differences. You may therefore need to adapt certain sections from my program if you wish to use them.

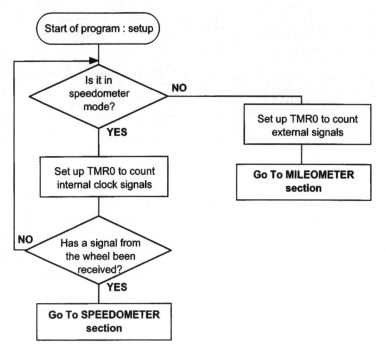

Figure 2.31a

SPEEDOMETER SECTION

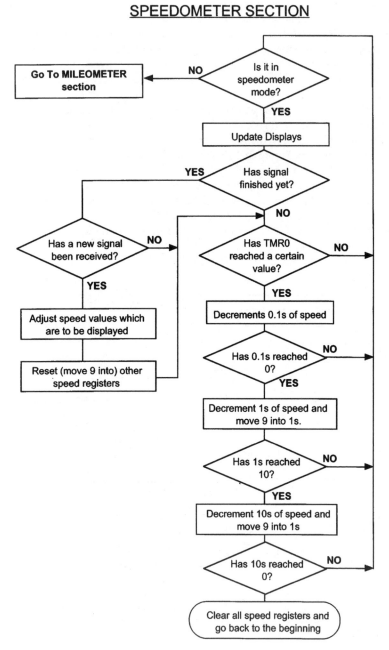

Figure 2.31b

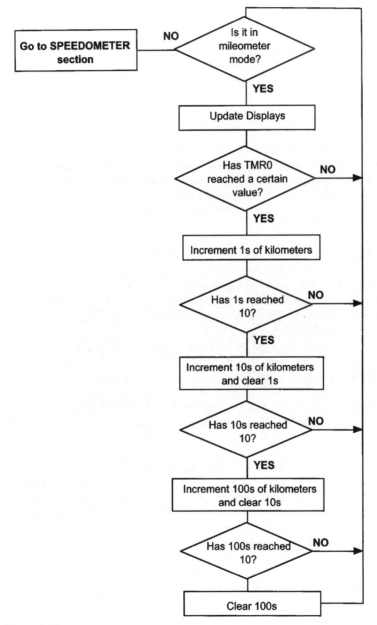

Figure 2.31c

3
Exploring the P16C71

Why use the P16C7x series?

Now that you are familiar with the PIC5x series, learning about other PICs is a lot more straightforward. New PICs are coming out all the time, and you need to be able to interpret the new options they offer. In general the instructions you use are identical, and the only difference is really the extra special function file registers the PIC offers. These form the basis for examining the difference between two PICs.

The 7x series offers a lot more tricks than the 5x series, but is therefore quite significantly more complicated. It is not recommended for beginners to use 7x PICs unless they need something particular to that chip (e.g. interrupts or analogue to digital conversion). Unlike the PIC54 and PIC55 where the only difference between each other's complexity lies solely in the extra input/output port, the PIC73 is considerably more complicated than the PIC71, and considerably less complicated than the PIC74.

Looking in particular at the pin layout of the PIC71 in Figure 3.1, you should notice that it is similar to the familiar PIC54. All the pins are in the same place making the two compatible as far as the circuit board is concerned; however we are given an extra I/O pin. If you are not using the timer zero clock input, you can use pin 3 as RA4 (bit 4 in Port A). Members of the PIC7x series therefore have a 5 bit Port A. You should also see some of the pins labelled AN0 and AN1, etc.; these can be made analogue inputs. There are up to 4 analogue inputs on the PIC71, 5 on the PIC73, and 8 on the PIC74. Finally you will notice a pin labelled Vref (pin 2 on the PIC71); this input can be made the voltage reference for the other analogue inputs (i.e. the PIC compares the voltage at the others with the voltage at the Vref pin). As you can imagine there are lots of combinations of the use of these pins, but you don't get complete flexibility on how they are arranged. You can't for example just make AN2 the only analogue input, and have the others digital. The precise details of your choices will be discussed in the appropriate section.

Finally you will notice the INT pin (which is also RB0), it can be set up so as to interrupt the program when it receives a particular signal. RB4-RB7 can also cause a different type of interrupt, though this isn't shown on Figure 3.1.

You will notice a great selection of other pin options on the more complex PICs; these will not be examined as the simpler PIC71 is really the subject of this chapter.

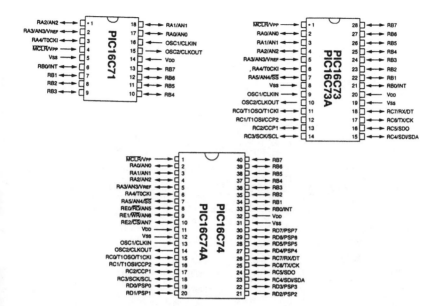

Figure 3.1

The inner differences: an extra bank

Having looked at the external differences between the 5x series and the PIC71, we now need to examine the inside of the PIC. Figure 3.2 should look vaguely familiar, and shows the arrangement of the file registers on the PIC71.

The primary difference is that there are two *banks*. Whereas the 5x PICs had only one bank ('filing cabinet'), the PIC71 has two sets of file registers. However, you should also notice that some are the same in bank 1 as in bank 0. Think of a bank as a state of consciousness for the PIC. When it is at one level, it views file registers one way, and when at another level it may (or may not) view the file register differently. File register 03 will always be the STATUS register, regardless of the bank (or frame of mind) the PIC is in. However in bank 0, file register 05 will be Port A, and in bank 1 file register 05 actually corresponds to a file register called TrisA. Even if I actually write Port A, it is translated to 05 by the assembler and then interpreted as Tris A, if the PIC is in the bank 1 'frame of mind'.

To switch the PIC from one bank to another we use one of the bits in the STATUS register (now you see why STATUS is the same in both banks – if it didn't exist in bank 1 there would be no way of getting back to bank 0!). This bit is called RP0 and is bit number 5. To go to bank 1, set the bit, and to return to bank 0, clear it.

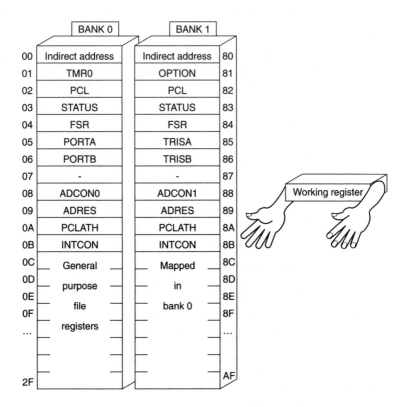

Figure 3.2

Example 3.1 You want to clear the file register called TrisA, however the PIC is currently in bank 0:

bsf	STATUS, 5	; goes to bank 1
clrf	TRISA	; clears the file register
bcf	STATUS, 5	; goes to bank 0

As both TRISA and PORTA are translated to the same thing (05) in the end, you could also write:

bsf	STATUS, 5	; goes to bank 1
clrf	porta	; clears the file register
bcf	STATUS, 5	; goes to bank 0

Naturally writing TRISA makes a lot more sense, but remember if you try to do something to Port A when in bank 1, you will actually do it to TrisA.

Each bank 0 file register is in some way related to its bank 1 equivalent (e.g. the OPTION register is the setup register for TMR0). Because the bank 1 file register tends to be involved in setting up, you generally only need to go into 'bank 1 mode' in the **Init** subroutine. Finally, please note that the PIC starts up in bank 0.

A different way of setting up

From the top, the first new file register we come across is the OPTION register. It isn't strictly a new file register because there was an OPTION register on the 5x PICs, however we did not have direct access to it. Remember how to move a number into the OPTION register (in order to set up TMR0) with the PIC5x? Below is a reminder:

```
movlw    b'xxxxxxxx'    ; moves the number into the w. reg.
option                  ; moves the w. reg. into OPTION
```

With the PIC7x series (and most others which follow) there is no need for the **option** instruction as we can simply move the number into the OPTION register.

```
[assuming the PIC is in bank 1...]
movlw    b'xxxxxxxx'     ; moves the number into the w. reg.
movwf    OPTION_REG ; moves the w. reg. into OPTION
```

First you should note that we use the term **OPTION_REG**; this is the way it is written in the lookup file, and has the '**_REG**' at the end to distinguish it from the instruction **option**. Secondly I should point out that if you don't switch to bank 1 before you do this, you will simply move the number into TMR0. There are two new additions to the OPTION register in the PIC71 which are found on bits 6 and 7 (previously non-assigned bits). If bit 7 is clear, all of Port B will have pull-ups enabled, and if set it won't (bit 6 will be explained in the interrupts section). Take care to remember to set this bit if you don't want this feature; it is very often forgotten. While on the subject of timing, please note the frequencies for the RC combinations shown in Table 3.1 which are different from those of the PIC5x.

Precisely the same applies to the **tris** instruction. With the 5x PICs you do the following to set up the ports inputs and outputs:

```
movlw    b'xxxx'        ; moves the number into the w. reg.
tris     porta          ; sets up Port A's inputs and outputs
```

Now we can forget about the **tris** instruction and write:

> **movlw**　　**b'xxxxx'**　　　**; moves the number into the w. reg.**
> **movwf**　　**TRISA**　　　　**; sets up Port A's inputs and outputs**

Again this has to take place when the PIC is in the bank 1 frame of mind. You may have noticed that a five bit number is moved into TRISA, this is because of the extra bit in Port A which we have in the PIC71. One of the bonuses of this system is that if all the bits of a port are to be outputs, we can set them up in one instruction:

> **clrf**　　　　**TRISA**　　　　**; moves 0 into TRISA, therefore all**
> **;　outputs**

This is naturally the same as:

> **movlw**　　**b'00000'**　　　**; moves 0 into the working register**
> **movwf**　　**TRISA**　　　　**; sets up Port A's inputs and outputs**

Table 3.1

Cext	Rext	Average Fosc @ 5V, 25°C	
20 pF	4.7 k	4.52 MHz	± 17.35%
	10 k	2.47 MHz	± 10.10%
	100 k	290.86 kHz	± 11.90%
100 pF	3.3 k	1.92 MHz	± 9.43%
	4.7 k	1.49 MHz	± 9.83%
	10 k	788.77 kHz	± 10.92%
	100 k	88.11 kHz	± 16.03%
300 pF	3.3 k	726.89 kHz	± 10.97%
	4.7 k	573.95 kHz	± 10.14%
	10 k	307.31 kHz	± 10.43%
	100 k	33.82 kHz	± 11.24%

New file registers

From file register 08 down, it will all look unfamiliar, and these will now be examined.

All but file register 08 are the same in both banks, and even then **ADCON1** is really just a two bit extension of **ADCON0**. These are used to set up the analogue to digital (A/D) conversions, and can be read to check out the conversion status.

ADRES, file register 09, contains the result of an A/D conversion, as a number between 0 and 255.

PCLATH contains bits 12 to 8 of the program counter, though they are not directly accessible to you. As a general point on the difference between the 5x and 7x, we see that where certain file registers were, in effect, hidden from us on the 5x PICs, they are out in the open on the 7x PICs.

Finally **INTCON** is used to set up the interrupts, and to interpret them. All this is explained in the interrupts section.

In the PIC73 and PIC74 there are up to 31 more of these special purpose file registers.

You are given 36 general purpose file registers to work with (0C to 2F) on the PIC71, and these are read the same in both banks. The PIC73 and PIC74 provide 192 general purpose file registers.

What are interrupts?

One of new 'tricks' on the 7x chips is the *interrupt*. This basically tells the PIC to drop whatever it's doing and go to a predefined place (the interrupt service routine), when a certain event takes place. This event could be receiving a signal on the INT (RB0) pin, or perhaps the state of one the bits RB4–RB7 changing. If TMR0 overflows (goes past 255 and resets to 0), the PIC can be made to interrupt, and it can also interrupt when an A/D conversion has finished. A/D conversions are explained in greater depth later on, but to understand this particular interrupt it is necessary to know that when you tell the PIC to perform an A/D conversion, it will take the PIC a short while to finish performing it. So after you begin the A/D conversion, the PIC will allow you to get on with other things, and will interrupt when the conversion is finished. Going back to the external interrupt (the INT pin), bit 6 of the OPTION register dictates whether the interrupt occurs on the rising or falling edge of the signal. If set, the interrupt will occur on the rising edge, and if clear, on the falling edge.

All four of these interrupts can be turned on or off individually, and more than one can operate at the same time. The obvious question is then, 'How can you tell what made the PIC interrupt when it did?' Fortunately each interrupt type also has a *flag* assigned to it which can be tested to see if that particular interrupt has occurred.

All but one of these bits are held in the **INTCON** file register (see next page):

Bit no.	7	6	5	4	3	2	1	0
Bit name	GIE	ADIE	T0IE	INTE	RBIE	T0IF	INTF	RBIF

RB change flag:
1: One of the bits RB4–RB7 have changed.
0: None of these bits have changed.
[**Note:** Must be cleared by you]

Ext. int. flag:
1: The ext. interrupt has occurred.
0: The ext. interrupt has not occurred.

TMR0 interrupt flag:
1: TMR0 has overflowed.
0: TMR0 hasn't overflowed.
[**Note:** Must be cleared by you]

RB change enable:
1: Enables RB change interrupt.
0: Disables it.

Ext. int. enable:
1: Enables ext. interrupt.
0: Disables it.

TMR0 interrupt enable:
1: Enables TMR0 interrupt.
0: Disables it.

A/D conversion interrupt enable:
1: Enables A/D interrupt.
0: Disables it.

Global interrupt enable:
1: Enables all interrupts which have been selected above.
0: Disables all interrupts.

Bits 0 to 2 are interrupt flags. These can be used in the following way (assuming this is in the place the PIC goes when an interrupt occurs):

```
btfsc     INTCON, 0    ; tests RB change flag
goto      RBChange     ; set, so RB change interrupt occurred
btfsc     INTCON, 1    ; tests external interrupt flag
goto      External     ; set, so external interrupt occurred
btfsc     INTCON, 2    ; tests TMR0 interrupt flag
goto      TMR          ; set, so TMR0 overflowed
goto      ADFinish     ; clear, so all that is left is the A/D int.
```

The places you go to in the above example are arbitrary; you are better off calling them something to do with what caused the interrupt, e.g. if a push button is connected to the external interrupt, give the label a name related to the push button, and its purpose. I have also assumed that if none of the first three interrupts have occurred, it is the completion of the A/D conversion that has caused the interrupt. There is an A/D interrupt flag, but as you can see it is not held in the INTCON register – it is bit 1 in the ADCON0 register (ADCON0, 1).

Three out of the four flags (all except the ext. int. flag) need you to actually clear them in order to reset them. This is simply done using the **bcf** instruction:

bcf   INTCON, 0   ; resets RBchange interrupt flag

Bits 3 to 6 of INTCON allow you to select which interrupts you want to be enabled (activated). Remember, you can change these anywhere in your program, as many times as you want.

Finally bit 7 is the global interrupt enable, which is the master switch for all these interrupts. Turn it off and no interrupts are enabled, but the state of the other enable bits in INTCON don't change. Turn it on and the interrupts which you selected in bits 3 to 6 become enabled.

It is generally best to set up INTCON during the **Init** subroutine, so that all setting up is in the same place.

The program has interrupted ... what next?

When the PIC is interrupted, it will go to the instruction at address **0004**. What's more, it actually calls a subroutine which starts at address **0004**, this is so that after dealing with the interrupt, the PIC can return to where it left off. Normally, address **0004** is five lines into the **Init** subroutine, so we will need to make some changes. What we really want the PIC to do is goto somewhere called the *interrupt service routine* (or **isr**). Once the PIC gets to the instruction at **0004**, it has already gone one level deep into the stack, so rather than calling the subroutine **isr**, we merely go to it. When the PIC comes across the return instruction in the **isr**, it will return to the line the PIC would have executed had the interrupt not have taken place, rather than to address **0004**. (**Important note:** the PIC7x series begin at the first address: **0**, rather than the last address: **1FF** like the

PIC5x series.) Your first idea on how to set this up may be putting the following at the very beginning of the program:

```
        org    0
        goto   Start

        org    4
        goto   isr

Init           etc.
```

The only problem with this is that you are wasting addresses **0001** to **0003**; however this is not serious. Alternatively you could write the following:

```
              org    0
              goto   Start

Init          clrf   porta          ;
              clrf   porta          ;
              goto   InitContinue   ; skips address 0004
              goto   isr            ; at address 0004, goes to isr
InitContinue  etc.                  ; carries on with rest of
                                    ;   subroutine
```

Counting down you should see that the line **goto   isr** is still at address **0004**, and rather than losing three lines (**0001** to **0003**), we really only waste one line (**0003**) – which was **goto   InitContinue**.

The interrupt service routine should begin by checking what event caused the interrupt (if more than interrupt is enabled), and then branch off to the relevant section in the isr.

Fortunately the PIC automatically clears the global enable bit in the INCON register upon an interrupt occurring. This means that no interrupt can take place in the interrupt service routine – you can imagine the havoc that would take place should this not be the case! Thus at the end of the interrupt service routine we would have to set the global enable just before returning, but even if we did this, an interrupt could take place immediately upon setting the global enable, and before actually returning from the isr. We would therefore have to set the global enable *after* returning, but we don't know where the PIC is going to return to, as we write the program. We actually set the global enable *at the same time* as returning, using the following new instruction:

```
        retfie                 ;
```

This **returns** from a subroutine and sets the global interrupt enable bit; it is naturally not available on the PIC5x series.

In certain cases you will want to return from the isr (or indeed any subroutine) *without* setting the global enable bit. In the past you will have used the **retlw** instruction, but we can now use the following:

> **return**   ;

This simply **return**s from a subroutine, and is not available on the 5x PICs.

If an interrupt occurs during sleep, the chip will wake up and do one of two things, depending on the state of the global enable bit. If the global enable is clear, the PIC will simply wake up and carry on from the line after the **sleep** instruction. If the global enable bit is set, the PIC will execute the instruction after **sleep,** and then call the interrupt service routine. Therefore, if you want the PIC to merely carry on from where it went to sleep, you must clear the global enable *before* the **sleep** instruction. If you then want the interrupts enabled, you should then set the global enable *after* the **sleep** instruction.

Example 3.1 Make the PIC go to sleep until TMR0 (which, incidentally, keeps on counting) overflows. It should then carry on with the rest of the program with the TMR0 and A/D conversion interrupts enabled:

```
movlw b'00100000'      ; only enables TMR0 interrupt
movwf INTCON           ;   (disables global)
sleep                  ; goes to sleep
movlw b'11100000'      ; enables TMR0, A/D, and global
movwf INTCON           ;   interrupts
```

Exercise 3.1 Write the *five* line program section to make the PIC go to sleep, be woken up by a change on either RB4, 5, 6 or 7. Upon waking up the PIC should *only call the interrupt service routine*, and then upon returning, enable the external interrupt. Do not write the interrupt service routine, but do set up INTCON before sending the PIC to sleep. (**Hint**: You should use the **nop** instruction.)

That's really all there is to interrupts; just remember to make the interrupt service routine fairly short, because you can't get an interrupt while you're in it. Think clearly when writing this part of the program – it may become quite complicated.

A new program template

With all these new file registers, it is clear that our program template needs to be updated. In it I have included A/D conversion file registers which we have not yet covered. They are all to do with setting up and thus have numbers moved into them in the **Init** subroutine. Remember also that ADCON1 is in bank 1.

```
;************************************
; written by:                       *
; date:                             *
; version:                          *
; file saved as:                    *
; for PIC...                        *
; clock frequency:                  *
;************************************

; PROGRAM FUNCTION:_____
;_____

          list    P=16C7x
          include "c:\pic\p16c71.inc"

;============
; Declarations:

          org     0
          goto    Start

;===========
; Subroutines:

Init          clrf    porta        ; resets input/output ports
              clrf    portb
              goto    InitContinue ; skips address 0004
              goto    isr          ; at address 0004, goes to isr
InitContinue
              bsf     STATUS, 5    ; selects bank 1
              movlw   b'xxxxx'     ; sets up which pins are inputs and
              movwf   TRISA        ;   which are outputs
              movlw   b'xxxxxxxx'  ; sets up which pins are inputs and
              movwf   TRISB        ;   which are outputs

              movlw   b'xxxxxxxx'  ; sets up ADCON1
              movwf   ADCON1       ;

              movlw   b'xxxxxxxx'  ; sets up timing register
              movwf   OPTION       ;

              bcf     STATUS, 5    ; goes back to bank 0
              movlw   b'xxxxxxxx'  ; sets up A/D register
              movwf   ADCON0       ;
```

```
        movlw   b'xxxxxxxx'   ; sets up interrupts register
        movwf   INTCON        ;

        return                ; returns from the subroutine
```

;==============
; Program Start:

Start
```
        call    Init
```
Main
```
        (Write your program here)
```
END

In most respects this is fairly similar to the previous program template. Remember to replace the line **list P=16C7x** with the actual PIC you are using, e.g. **P=16C71** in the case of the PIC71.

In the vast majority of cases, interrupts are a commodity rather than a necessity, so you may notice that the next project could in theory be constructed using the PIC54. The reason for using a PIC71 will more often than not be for its A/D conversion, and the fact that it has interrupts is normally just a bonus. In this project there is, as I hope you will appreciate, an acute difference between the two methods discussed, but the choice of the PIC71 in this case really comes down to demonstrating interrupts.

The project we will undertake to practise interrupts will be a quiz game device. There will be four push buttons (one for each player), four LEDs (one by each button to indicate which player pressed their button first), and a buzzer to show that a button has been pressed. There will also be a push button for the quiz master to reset the system, once the faster player has been spotted. This reset button need not actually be assigned an input on the PIC, and can simply be connected to the MCLR pin. You may wonder why we are going to the trouble of using interrupts for this project, which looks as if it may well be viable on the PIC54. However with the PIC54, the obvious method of testing to see which player presses their button first is through testing each button, one after another, and then looping back. Let us say, for example, that the PIC has just finished testing the first button, and then immediately afterwards the first button is pressed. The PIC then tests the second and third buttons, after which the fourth player responds. The fourth player's button is now tested, and as far as the PIC is concerned, the fourth player responded first. The times we are dealing with are, as you know, hundreds of thousandths of a second, but if we want to be really exact we can use interrupts. The four players' buttons would be connected to RB4 to RB7, and so the RBChange interrupt could be triggered by any one of them being pressed. We could then immediately swap the upper and lower nibbles of Port B to turn the appropriate LED on. For this to be possible the LEDs would have be on RB0 to RB3. You can see how it is often convenient to think about how the program is roughly going to work, while you assign the inputs and outputs. The buzzer will be connected to RA0. The circuit diagram is shown in Figure 3.3, and then the flowchart in Figure 3.4.

As you can see, the main body of the program is really nothing at all, just a constant loop. All the clever stuff happens in the interrupt service routine.

Exercise 3.2 Write the **Init** subroutine for this program. Remember, only the RBChange interrupt should be enabled, and don't forget to enable the global interrupt bit. Don't forget that address **0004** must be **goto   isr**, and that TMR0 will need to be prescaled at its maximum. Finally, think carefully about how you handle the bank 1 registers.

The main body of the program is just a loop, waiting for the RBChange interrupt to occur. The program, from **Start**, would therefore be:

| **Start** | **call** | **Init** | **; sets everything up** |
| **Main** | **goto** | **Main** | **; waits for RBChange interrupt** |

We now come to the **isr.** At the moment we are only using the RBChange interrupt, so it seems that we don't need to check any of the interrupt flags, but this may need to be changed. The RBChange flag should be cleared.

Because we have been careful with our choice of inputs and outputs, we can simply swap the upper and lower nibbles of Port B to get the appropriate LED to turn on, and then turning on the buzzer is also straightforward.

Exercise 3.3 What *two* lines will accomplish these two tasks.

The next task is waiting one second before turning off the buzzer. We could simply throw in the one second delay that you are familiar with, but instead we are going to use the TMR0 interrupt. If we are using a 2.47 MHz RC oscillator, instructions are executed at a frequency of 617 500 Hz (2 470 000/4), and TMR0 counts up at a frequency of 2412 Hz (617 500/256). Assuming TMR0 starts from 0, the frequency of the TMR0 interrupt is 9.42 Hz (2412/256). We can therefore count about 10 TMR0 interrupts after which 1.06 seconds (10/9.42) will have passed. To be able to use the TMR0 interrupt we will first have to enable it, and then disable the RBChange interrupt so that it doesn't respond to any button being pressed or released. We must, however, leave the global interrupt disabled until we return from the isr. Finally, as we are using two interrupts, we will have to distinguish between them at the beginning of the isr (see page 113):

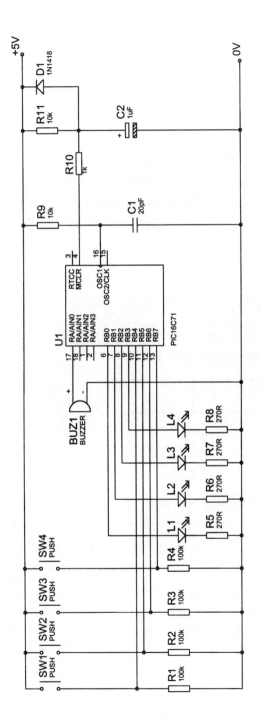

Figure 3.3

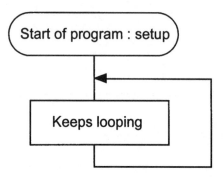

ISR ...

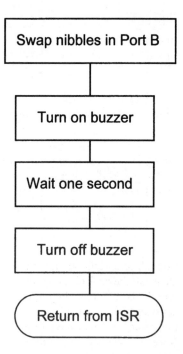

Figure 3.4

```
isr       btfss        INTCON, 0  ; tests RBChange flag
          b            Timer      ; TMR0 interrupt occurred
          [swap Port B etc.]      ; RBChange interrupt occurred
```

After the line where the buzzer is turned on, a new number must be moved into INTCON to enable the TMR0 interrupt and disable the RBChange interrupt. After this the PIC should return, while at the same time enable the global interrupt bit.

Exercise 3.4 What *three* lines achieve this?

The PIC will then keep looping until the TMR0 interrupt occurs (admittedly this first timing varies considerably, but the next nine will be exact, creating a total time of roughly one second). The isr jumps to a section called **Timer** when the TMR0 interrupt occurs. First we need to clear the TMR0 interrupt flag, and then we should wait for the TMR0 interrupt to happen 10 times before turning off the buzzer. We will therefore have a general purpose file register which holds the number 10 to start with, and which is decremented every time the TMR0 interrupt occurs. After the tenth time, the buzzer is turned off and the PIC goes to sleep (the lower power consuming mode). When the PIC goes to sleep the state of its outputs stays the same, so the correct LED stays on. Before going to sleep, the PIC should set up INTCON so that the RBChange and global interrupts are enabled. The section called **Timer** should therefore be the following:

```
Timer    bcf          INTCON, 2 ; resets TMR0 interrupt flag
         decfsz       Ten, f    ; has this happened 10 times?
         retfie                 ; no, so returns, enabling the global
                                ;   int. bit
         bcf          porta, 0  ; yes, so turns off buzzer
         movlw        b'10001000' ; global and RBChange are enabled
         movwf        INTCON    ;
         sleep                  ; goes into low power consuming
                                ;   mode
         return
```

When the PIC wakes up from sleep and the global interrupt is enabled, it will execute the next line, and then call the isr. This is convenient as it gives the PIC an opportunity to return from the isr before calling it, thus avoiding any potential problems involving overloading the stack.

In the **Init** subroutine the number 10 will have to be moved into **Ten**, and don't forget to define it. (**Note:** The general purpose file registers start from 0C on the PIC71 rather than 08 as on the PIC5x series). The whole program is shown in Program M in Chapter 7.

We have looked at two interrupts, the RBChange and TMR0 ones. The TMR0 interrupt can be very useful in performing timings which needn't be in exacts seconds; as you may remember we could only cause a 1.06 second delay rather than a 1 second delay. Obviously we can't use the A/D interrupt until we have mastered A/D conversion, and we will practise the external interrupt later.

Analogue to digital conversion: what is it?

The great advantage of the 7x series over the 5x series is the analogue to digital conversion it offers. This means that the PIC is able to read the voltage at one of its four analogue inputs, and convert this reading into a number between 0 and 255. This provides the user with a unit of 0.02V when using a 5V supply. For example, if the result of an A/D conversion was 10, the input voltage would be about 0.2V, and if it was 200, the input would at be about 4V. This allows much greater flexibility than the 5x series which can only tell whether an input is high or low (more than 2.5V or less than 2.5V respectively). (**Note**: The voltage it reads is compared to V$_{DD}$ (the supply voltage), but you can make the chip compare the voltage to that of another input by making RA3 the Vref. You can thus see whether two inputs stay the same, or by how much they differ.)

A/D conversion can be a fairly lengthy process (when compared to the speed with which most instructions are executed), and so when you begin an A/D conversion it may be a while before a result is obtained and stored in the file register ADRES. The time an A/D conversion takes can be changed by you; the longer you let it take, the more accurate the result. This is just part of the important setting up process, necessary before any conversions take place. This setting up should be placed in the **Init** subroutine, moving certain numbers in the file registers ADCON0 and ADCON1.

ADCON0

Bit no.	7	6	5	4	3	2	1	0
Bit name	ADCS1	ADCS0	-	CHS1	CHS0	GO	ADIF	ADON

A/D on bit:
1: A/D converter is on.
0: A/D converter is off (and consumes no operating current).

A/D int. flag:
1: The A/D interrupt has occurred.
0: The A/D interrupt has not occurred. Must be cleared by you.

GO/DONE:
1: Conversion is in progress. Setting this bit starts conversion.
0: Conversion has finished.

Channel select: Selects which input to read.
00: RA0/AN0
01: RA1/AN1
10: RA2/AN2
11: RA3/AN3

Reserved: Do what you want with it!

A/D clock select: Selects how long the PIC takes over a A/D conversion
00: Fosc/2
01: Fosc/8
10: Fosc/32
11: FRC

Bit 0 is simply the on/off switch for the A/D converter. When it is set, the A/D converter is on and consuming extra current, clear the bit and it switches off. Bit 1, the A/D interrupt flag, is like all the other interrupt flags found in INTCON, and must be cleared in your program after an interrupt occurs. Bit 2 is the bit you set to begin an A/D conversion, and the one you test to see when it has finished (should you decide not to use the A/D interrupt). If you're not using the interrupt, and don't want anything to happen while you're waiting for the conversion to take place, this is what you might use:

```
          bsf     ADCON0, 2    ; starts A/D conversion
ADLoop    btfsc   ADCON0, 2    ; has conversion finished?
          goto    ADLoop       ; no, so keep looping
          etc.                 ; yes, so exits loop
```

Bits 3 and 4 together select which analogue input you want the PIC to examine. If, for example, you wished to test the voltage at AN2, bit 3 would be 0, and bit 4 would be 1. You would then have to start the A/D conversion etc., but this tells the PIC which input you are talking about. Bit 5 is reserved, this means that you can do anything you want with it, it is in effect a general purpose bit. You are, however, advised to leave it alone in order to maintain upward compatibility (i.e able to use your program in later PICs, which may have a specific use for this bit).

Finally, bits 6 and 7 together select the time the PIC takes over an A/D conversion. If you make it too fast, the conversion will be inaccurate, make it too slow and it will (by definition) take a long time. Fosc/2 means that the PIC divides the frequency of the oscillator which you have connected to it by 2.

Example 3.3 You have connected a 4 MHz oscillator to the PIC, and have set the A/D clock cycle to Fosc/2. The frequency for the A/D clock is therefore 2 MHz (= 4/2), and one cycle lasts 0.5 microseconds (seconds $\times$ 10^{-6}) = 1/2 000 000. 0.5 µs is too little time, something between 2 and 8 microseconds is recommended, but naturally you can make it longer if you want.

Similarly Fosc/8 divides the external oscillator frequency by 8, and Fosc/32 divides the frequency by 32. Alternatively, if you want the A/D conversion clock to be independent of the external oscillator, select the FRC option. It gives a typical conversion time of 4 microseconds. Table 3.2 summarizes the conversion times for certain settings:

Table 3.2

AD clock source		Device frequency				
Operation	ADC 1:0	20 MHz	16 MHz	4 MHz	1 MHz	333.3 kHz
2Tosc	00	100ns	125ns	500ns	2.0µs	6µs
8Tosc	01	400ns	500ns	2.0µs	8.0µs	24µs
32Tosc	10	2.0µs	2.0µs	8.0µs	32µs	96µs
RC	11		2–6µs	2–6µs	2–6µs	2–6µs

ADCON1

ADCON1, the bank 1 version of ADCON0, selects which Port A inputs are analogue, and which are digital. It is also used to choose the voltage reference (supply voltage or RA3). Because ADCON1 is made up of only two bits, you are only given the four possible combinations shown in Table 3.3.

Table 3.3

Bit 1	Bit 0	RA0	RA1	RA2	RA3	Ref
0	0	A	A	A	A	VDD
0	1	A	A	A	Vref	RA3
1	0	A	A	D	D	VDD
1	1	D	D	D	D	VDD

Once the A/D conversion has finished, the result is stored in ADRES, replacing the previous value. The value in ADRES may then be moved into another file register, or tested directly (by subtracting a number from it and testing to see whether the result is positive, zero or negative). This can be done using the zero and carry flags. The zero flag will be set when the result is zero and clear if the result isn't zero. The carry flag (after a subtraction) is set if the result is positive or zero, and clear if the result is negative. A combination of testing these bits will allow you to interpret the analogue result practically in every possible way.

To practise A/D conversion, our next project will be a temperature sensing device which indicates whether the temperature of your bath is too high or too low. There will be three LEDs to indicate whether the temperature is below, inside or above the suitable temperature range. These will be connected to RB0 to RB2. RA0 will be the analogue input connected to a temperature sensor LM35 which varies its output linearly according to temperature. The circuit diagram is shown in Figure 3.5, and the flowchart in Figure 3.6.

As with the previous program, the actual main loop of the program is practically nothing at all. In this case we simply need to keep starting A/D conversions. The program from **Start** would therefore be:

```
Start      call    Init        ; sets everything up
Main       bsf     ADCON0      ; starts A/D conversion
           goto    Main        ;
```

Looking at the above program segment, we can see that having an A/D interrupt isn't actually necessary. We could simply have the following arrangement:

```
Start      call    Init        ; sets everything up
Main       bsf     ADCON0      ; starts A/D conversion
ADLoop     btfsc   ADCON0, 2   ; has conversion finished?
           goto    ADLoop      ; no, so keep looping
           etc.                ; yes, so exits loop
```

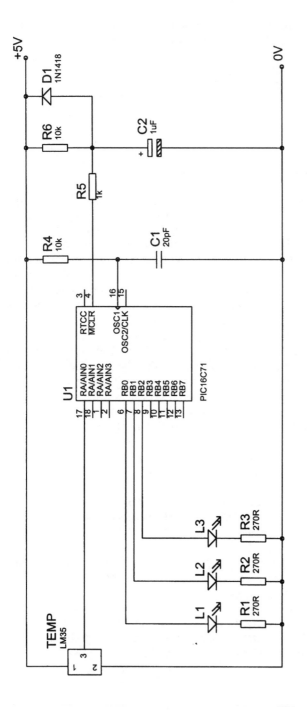

Figure 3.5

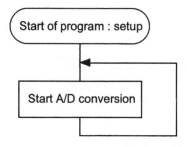

ISR ...

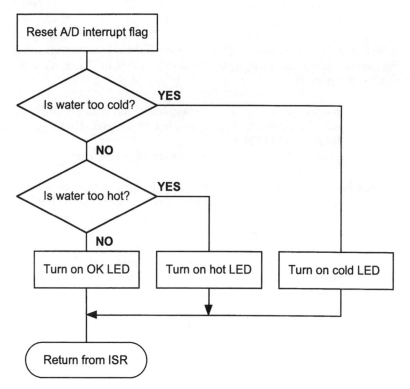

Figure 3.6

However, in a more advanced version of this program we may want to have a more complex main program loop, and it such a case the A/D interrupt would be very useful. We will therefore keep using the A/D conversion interrupt.

Exercise 3.5 Write the **Init** subroutine for this program.

In the isr, we needn't test the A/D interrupt flag, as it is the only interrupt which has been enabled, but we do need to reset it.

Exercise 3.6 What line will reset the A/D interrupt flag?

Following the flowchart, we see that the next step is to see whether the temperature is too cold (i.e. whether the analogue result is below a certain value). We do this by subtracting the lower limit from the analogue result (held in the file register ADRES), and testing the carry flag. If the number in ADRES is less than the lower limit, the result of the subtraction will be negative and the PIC should branch to a section labelled **Cold**. This particular temperature sensor gives an output voltage of 0.01V per degree Celsius. If we say that the minimum bath water temperature is 35°C, then the minimum input voltage is 0.35V. We then divide 0.35 by 5 leaving 0.07, and 0.07 256 = 18. The minimum A/D result is therefore 18:

```
movlw   d'18'          ; compares result with the decimal
subwf   ADRES, w       ; number 18 without affecting ADRES
btfss   STATUS, C      ;
goto    Cold           ; less than 18, so too cold
```

The maximum temperature shall be, for example, 45°C. Using the same technique described previously we can work out that the maximum A/D result is 23. If the result is less than 23 the isr should branch to a section called **Ok**, and if it isn't it should branch to **Hot**:

```
movlw   d'23'          ; compares result with the decimal
subwf   ADRES, w       ; number 23 without affecting ADRES
btfss   STATUS, C      ;
goto    Ok             ; less than 23, so OK
goto    Hot            ; more than 23, so hot
```

The **Cold** section should turn on only the cold LED by moving a specific number into Port A. Afterwards the PIC should return, at the same time enabling the global interrupt. Likewise the **Ok** and **Hot** sections should perform similar tasks except with different LEDs.

Exercise 3.7 Write the **Cold**, **Ok**, and **Hot** sections. They should consist of *three* lines each.

The entire program has now been completed and is shown as Program N in Chapter 7.

Some final points

Although the two previous examples have not involved interrupting in the middle of a proper program loop, there will be many cases where an interrupt may occur just at the moment when an important number has been moved into the working register, or perhaps the STATUS register holds an important result. These two file registers are likely to be changed during the isr, and so it may be important to save the numbers in these at the very beginning of the isr, and then reload them with the numbers at the end. The instructions involved would be:

```
movwf   wtemp          ; moves working register into the GPF
                       ;  wtemp
movfw   STATUS         ; moves STATUS register into the
movwf   statustemp     ;  GPF statustemp, via the working
                       ;  register
```

[The isr...]

```
movfw   statustemp     ; moves statustemp into the STATUS
movwf   STATUS         ;   register via the working register
movfw   wtemp          ; moves wtemp into the working
                           register
```
[return from isr]

It is likely that you will be returning from the isr from more than one place, so rather than copying out the reloading instruction set in every place you want to return from, it is a good idea to branch to a pre-return section wherever you would have returned. This pre-return section would resemble the following:

```
prereturn movfw  statustemp   ; moves statustemp into the STATUS
          movwf  STATUS       ; register viathe working register
          movfw  wtemp        ; moves wtemp into the working
                              ;   register
          retfie or return    ; returns from isr
```

Throughout the isr, where you would have written **retfie**, write **goto   prereturn** instead.

Note also that the PIC7x has a stack which is eight levels deep. This means that you can call a subroutine within a subroutine within a subroutine within a subroutine within a subroutine within a subroutine within a subroutine within a subroutine! This is quite useful – the third level in particular as the others are not used that often.

Finally, we have been given two more instructions:

addlw   number   ;

(Not for PIC16C5x series) – **add**s a **number** with the number in the working register.

sublw   number

(Not for PIC16C5x series) – **sub**tracts the number in the working register from a **number**.

Final PIC71 program

This final PIC71 project will tie together all of the ideas we have studied so far – it will be the most complicated, but probably also the most useful. I would like to thank Max Horsey who came up with the idea and designed the circuit, for the its use in this book. It will be an multipurpose random number generator. Its five modes will produce random numbers between 1 and 6, 1 and 12, 1 and 12 (with probabilities such that it is comparable to using two dice), and 1 and 99. There will also be a National Lottery number generator, producing six different numbers between 1 and 49. It will automatically switch off if not used after 3.5 minutes.

It will have four seven-segment displays, to show the number, a push button to select the mode and two contacts to be pressed when a number is required. The two contacts are in effect a 'vibe' sensor. By holding your finger over the two contacts you make the connection using the resistance of your skin. This resistance is measured (using the analogue input) and used to generate the random number. In this way it produces personal random numbers (particularly important when playing the National Lottery).

The four seven-segment displays can be strobed, so that they only require 11 outputs, the push button will require one input, and the contacts an analogue input, creating a total of 13 I/O pins. It will therefore fit on the PIC71. The push button will be connected to RB0 as an external interrupt, the controlling pins of the seven-segment displays to RA1 to RA4, and the seven display pins from *both* of the displays to RB1 to RB7. The contacts will use the AN0 (RA0). All this is shown in the circuit diagram (Figure 3.8).

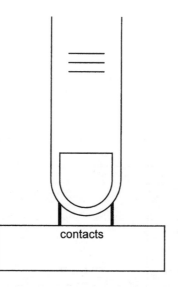

contacts

Figure 3.7

The next step is to construct a flowchart. This will be a very complex program, so a good flowchart is really vital. We won't be able to foresee all the complications from this early stage, so as with most complicated programs, we will have to add some things as we go along. The basic flowchart is shown in Figure 3.9, but there are many complications not shown in the flowchart.

One of the first things the PIC will need to do is work out which mode it should be in, and display this on the seven-segment displays. The mode will be shown as: **1-6_**, **1-12**, **-6-6**, **1-99** and **Lott** ('_' denotes a blank space). When the device first starts up, it should be in the 1-6 mode.

Lots of different things will be going in the program and the display subroutine will have to be regularly called, if the strobing is to work properly. It will need to display the mode which the device is in, and then later the random numbers etc., so our conventional display method may have to be changed slightly. We will have four file registers called **dig1**, **dig2**, **dig3** and **dig4**. The display subroutine will display whatever code is in these through the correct seven-segment display, so no seven-segment encoding takes place in the display subroutine. Those four file registers act as a transporter for displaying code. In a section of the program, a number may be moved into one of them, and then when that particular display is next turned on, the effect will be noticed (this happens effectively instantly due to the PIC's high processing speed). Figure 3.10 shows how the four file registers are allocated:

When the mode push button is pressed it should trigger the external interrupt and cause the PIC to enter the isr. Once in the isr we can make the PIC branch to a mode changing section. This section will allow the device to scroll through the different operating modes, displaying everything on the displays. We need

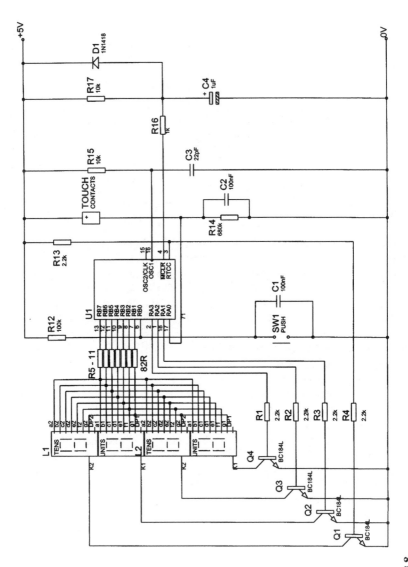

Figure 3.8

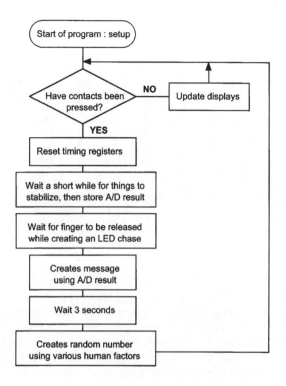

ISR ...

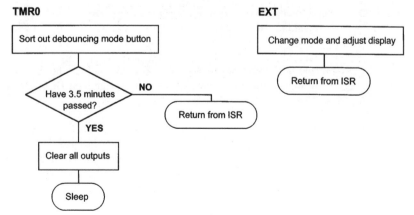

Figure 3.9

to debounce the button as well as waiting for it to be released (without stopping the displays from working).

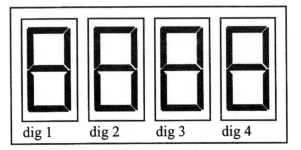

Figure 3.10

Exercise 3.8 Write the **Init** subroutine, don't forget to set up TMR0, prescaled at 256.

Our next step will be writing the external interrupt part of the isr. We will need some sort of general purpose file register to keep track on which mode the PIC is in. Each mode is thus assigned a number between 0 and 4. This file register (let's call it **Chooser**) will therefore need to be incremented when the external interrupt is triggered, thus selecting the next mode. If the number in Chooser gets to 5, we know that it has gone too far and must be reset to 0.

Exercise 3.9 Write the *five* lines which will achieve this.

The PIC will need to change the display every time a new mode is selected, so the PIC will then need to use the number in Chooser to move the correct codes into **dig1** to **dig4**. This can be done using the program counter.

Exercise 3.10 Write the *seven* lines that use the program counter and the number in Chooser to jump to one of five different sections labelled **Dis6**, **Dis66**, **Dis12**, **Dis99** and **DisLot**.

As three out of the five modes have the same first two symbols, they can all go to the same place when setting those first two displays up. In **Dis6**, **Dis12** and **Dis99** the PIC need only set up **dig3** and **dig4** individually. The arrangement of the seven segments on Port B is simply: a, b, c, d, e, f, g, -. To display the number six, the following number is therefore used: **10111110** (work it out!).

Exercise 3.11 Write the *five* lines in **Dis6** to move the correct numbers into **dig3** and **dig4**, and then jump to a section called **DisCommon**. Then do the same for **Dis12** and **Dis99**.

When returning from the isr, we do not want to enable the global interrupt bit, because we will need to debounce the mode button. This is done by leaving the interrupt disabled until it has been released for a tenth of a second.

Exercise 3.12 Write the *five* lines in **DisCommon**, which set up displays 1 and 2, and then returns from the isr (*not* enabling the global interrupt).

DisLot and **Dis66** will move the correct numbers into **dig1-4**, and then return.

Exercise 3.13 Write the *nine* lines which make up **DisLot**, and the *seven* which make up **Dis66** (think carefully on how to get it down to seven lines).

Thus the entire mode changing section is complete, however, as the PIC needs to start with **1-6** on the displays, we need to fake an external interrupt when it first starts up. This is simply done by setting the external interrupt flag and calling the isr (with the **call** instruction). Setting the flag tells the isr to go to the external interrupt section (rather than the TMR0 section). If the PIC is to start in 1-6 mode, the number 4 should be moved into **Chooser** in the **Init** subroutine. Thus when **Chooser** is incremented it will become 5, which is too big, and thus reset to 0 which is the mode number for 1-6.

Exercise 3.14 Write the *three* lines which should be executed before the PIC enters the main loop. These will set things up, and cause a fake external interrupt. Add the extra two lines to your **Init** subroutine which set **Chooser** up correctly (not in Answer section).

The PIC should then test the contacts, while still displaying everything. As they are connected to an A/D converter we need to see whether the A/D conversion result is over a certain minimum value. This has been found, through experimentation, to be 10. The PIC should first start an A/D conversion, and then call the **display** subroutine which handles the strobing of the four displays. The A/D clock is set to one eighth the operating frequency. An instruction is executed in one quarter of the operating frequency, so after two instruction the A/D conversion will have taken place. The **display** subroutine will have taken well over two instructions, and so by the time it returns the A/D conversion will have finished. There appears, therefore, to be little need for an A/D interrupt in this case. After calling the **display** subroutine the PIC, in order to debounce the mode button, should test to see whether it has been released. If it has been released the PIC should enable a TMR0 interrupt (and the global interrupt). When the TMR0 interrupt later occurs, the external interrupt can then be re-enabled, and the mode button will again affect the program.

Exercise 3.15 What should the first *six* lines of the **Main** loop be, if they are to start an A/D conversion, call the **display** subroutine, and then handle the debouncing of the mode push button. (**Note:** If the push button is still pressed, the PIC should jump to a section labelled **ADTest**.)

The PIC should now double check that the A/D conversion is now complete by testing bit 2 of **ADCON0**. Going back to **Main** and skipping one instruction if it is still in progress. Then we need to see whether the A/D conversion result is 10 or greater. If the result is less than 10, the PIC should go back to main.

Exercise 3.16 Which *six* lines will achieve this?

Once pressed, the PIC should reset the registers it is using to time the 3.5 minutes (until automatic switch off), so that it starts counting afresh.

Before we can reset the timing registers, we need to know the exact nature of the timing. We will therefore now write the section which will send the PIC to sleep after 3.5 minutes of the contacts not being pressed. Rather than constantly calling a subroutine and checking how much time has passed, we can use the TMR0 interrupt to make the PIC automatically jump to a certain section after a certain time has passed. The TMR0 interrupt is enabled whenever the mode changing push button isn't pressed (which is most of the time), so can be used for this purpose. Using a 2.47 MHz oscillator, and prescaling TMR0 by 256, the TMR0 interrupt occurs every 0.106 seconds. To wait 3.5 minutes we need the TMR0 interrupt to occur 1981 times. We can do this using two prescalers – one at 256 (the maximum), and another at 7. Though this doesn't produce a time of *exactly* 3.5 minutes, it doesn't matter for our purposes. This first thing we will need to do in the isr is test the external interrupt flag to see what event caused an interrupt. If the external interrupt flag is clear, the PIC should jump to **TMRInt**, otherwise it should continue with the mode changing section that you have already written.

Exercise 3.17 Write the *two* lines which will do this.

The first thing to happen in **TMRInt** is the enabling of the external interrupt (as by this time it will have been released for long enough to overcome button bounce). We can then begin to time 3.5 minutes – first we check whether this interrupt has occurred 256 times (by decrementing a file register which initially was clear, and testing to see if the result is 0). If it hasn't happened 256 times, the PIC should return while enabling the global interrupt. If it has happened, the PIC needs to test to see whether the TMR0 interrupt has occurred 256 times, 7 times. A similar technique is used. If the full 3.5 minutes have passed, the PIC should reset the second prescaler, clear both ports (so everything turns off), and go to sleep. Because we have enabled the external interrupt, pressing the mode changing push button will wake the PIC from sleep, but as the global interrupt isn't enabled the isr will (fortunately) not be called and the mode will not be changed. The PIC will simply carry on where it left off, so the instruction after **sleep** should make the PIC return from the isr, while enabling the global interrupt.

Exercise 3.18 Which *eleven* lines make up the **TMRInt** section? You will need two general purpose file registers as postscalers (call them **Post256** and **Post7** if you want).

Going back to the main loop, we now know how to reset the timing registers when the contacts are pressed. We should clear the first postscaler, and move seven into the second one. The PIC should also ignore the mode changing button (i.e. all interrupts) from then on.

Exercise 3.19 Which *four* lines will perform these tasks? These are to be performed once the contacts have been touched.

The PIC should then wait a very short while for the circuit created through the skin to stabilize, before storing the A/D conversion result. This can be done by performing the A/D conversion 256 times. The following section (you can call it **Stabilizer**) will start A/D conversion, and keep looping until it has finished. It should then wait for this to happen 256 times (you should know how to do this), looping back until it has. The PIC should then store the A/D result in a file register (call it **Skin** if you want).

Exercise 3.20 Write the *seven* lines which will achieve this (I called the file register which counts 256 times **ADCount**).

The PIC now needs to wait for the finger to be removed so as to break the circuit until this happens, the displays need to be chasing. We should first start an A/D conversion, and then sort out the chasing of the display (by calling a special subroutine - let's call it **Chaser**). This routine will change the numbers in **dig1** to **dig4**, and so we then need to call the **display** subroutine to keep the displays on while the contacts are pressed. After this, the PIC should check that the A/D conversion has finished, looping back to the beginning of this section (**Main2**), skipping the next instruction which started the conversion.

Exercise 3.21 Which *five* lines will achieve this? (Do not worry about the **display** or **Chaser** subroutines.)

Once it has been confirmed that the A/D conversion is finished, the result should be compared with 10, to see whether or not the contacts have been released. If they are pressed, the PIC should loop back to **Main2** and start another A/D conversion. Alternatively the PIC should wait for the result to become suitably low, 200 consecutive times (the resistance between the contacts is very variable and this confirms that the finger has really been removed). After the result has been confirmed, the general purpose file register used to count 200 times needs to be reset.

Exercise 3.22 Which *eight* lines will perform these tasks?

The PIC now needs some way of generating a random number (this is one of the hardest things for machines to do). Humans are much more random than PICs, so the best random number generator is linked to the human user. TMR0 can be made to count up 614 400 times a second, so if we look at the number in TMR0 at the moment the contacts are released, it will be quite random. In this way the human user dictates the number in TMR0, but has no real control over it. This random number will be between 0 and 255, and we need to squeeze this down to between 1 and 6 or 1 and 99, etc., but we will tackle this at a later stage. The PIC nevertheless needs to store the number currently in TMR0 in a general purpose file register.

Exercise 3.33 Which *two* lines will store the value? You could call the GPF **TouchTime**.

The PIC should now use the value in **Skin** to choose a suitable message (this can be done in a subroutine), and then wait 3 seconds. We could copy out the routine to create a 3 second delay, or could try and use the TMR0 interrupt section. The first prescaler in the **TMRInt** section extends the time length to about 27 seconds, and then the second one to the full 3.5 seconds. By changing the values of these prescalers we can still create the same delay in the **TMRInt** section, using different numbers. If we make the first prescaler create a delay of 3 seconds, we can then use a larger second prescaler to bring that length up to 3.5 minutes. 3/0.106 = 28.3, so if the first prescaler is 28, it will create a roughly 3 second delay. 3.5 minutes is 210 seconds, so the second prescaler should be 70. We will need to now go back and make these changes throughout the program it is normally useful to use Notepad to *Search* for the words **Post256** and **Post7** to make sure nothing is left out. You should naturally change the names of the postscalers as well.

To now use the TMR0 interrupt section to create a 3 second delay, we first need to reset the first postscaler, and then enable the TMR0 (and global) interrupts. We should then set a general purpose bit (e.g. **sec3**) which will be cleared in **TMRInt** when the first prescaler has reached 0, and keep testing it until it is cleared. Assign the **sec3** to a file register you laid aside for general purpose bits using the **#define** instruction in the declarations section. Don't forget to insert the line to clear **sec3** in the appropriate place in **TMRInt**.

Exercise 3.24 Write the *nine* lines which call the message choosing subroutine and then wait 3 seconds (while constantly calling the **display** subroutine). The loop involved may be called **Loop3Sec**.

Exercise 3.25 Then, in the **TMRInt** section, after the following pair of lines:

decfsz   Post28
retfie

... *three* lines must be added. The first two would reset the first postscaler, and the third should clear **sec3**. Write these *three* lines.

To get our random number we need to use the value in **TouchTime** (completely variable), together with the value in **Skin** (quite variable), to create a random number. We will therefore add the two together and store the result in a general purpose file register (I called it **Random**). The PIC then needs to use the value in **Chooser** (which tells it which mode it's in), to jump to one of five different places, where it can convert the number between 0 and 255 into something more suitable for that particular mode. These five sections could be called: **Ran6**, **Ran66**, **Ran12**, **Ran99** and **RanLot**.

Exercise 3.26 Write the *ten* lines that will achieve all this.

To change a number between 0 and 255 into one between 1 and 6, we need to repeatedly add the number 6 to the number between 0 and 255. Take the number 156 for example – add 6 to it and you get 162. Do this repeatedly until the number overflows past 255 and you get a number between 1 and 6: 162 + 6 + 6 ... + 6 = 252. 252 + 6 = 258 = 2. In this way we have converted the number 156 into 2. We can do the same for the other modes by adding 12, 99 or 49 repeatedly. One of the problems we face with this method is that adding 6 to 250 sets the carry flag, as the result is 0, however 0 is of no use to us. It would be far easier if we were looking for a number between 0 and 5, so we *will* look for a number between 0 and 5, and then add one to it afterwards. As the 1–6, 1–12, and 1–99 modes are very similar, we can use the same section for all three. For this to be possible we should add the number in a file register (**Scaler**), rather than the number itself. In this way we can move the appropriate number into the file register and then go to a section common to all three.

Exercise 3.27 Write the **Ran6**, **Ran12** and **Ran99** sections which move the appropriate number into the working register, and then branch to the common section (you may call it **Adder**). They should therefore consist of *two* lines each.

The section **Adder** should firstly store the number in the working register (I called the GPF **Scaler**). The number in **Scaler** should then be added to **Random**, and then the carry flag should be tested. If set, the number has grown past 255 and can be used, however if clear the PIC should loop back to **Adder**, skipping the instruction which moves the working register into **Scaler**. Finally, once out of the loop the PIC should add 1 to **Random** so that it becomes a number between 1 and 6 or 1 and 12, etc.

Exercise 3.28 Which *six* lines will perform these tasks?

We have completed three out of the five different random number generators. The double 1–6 generator should be completed by having two 1–6 generators. We can simply use the one we used for the 1–6 mode to begin with, so change the line:

b	**Ran66**

... to: **b**   **Ran6**.

At the end of the **Adder** section you should test bit 2 of the **Chooser** register which will be set if in **-6-6** mode. If clear the PIC should go to some other place – just write the **goto** (or **b**) instruction with no destination for the moment. Otherwise the PIC should skip that line and move **Random** into a general purpose file register which we shall call **Tens**. It should also get a new random number (by adding **TouchTime** to **Random**), and convert it into something between 1 and 6. After this the PIC should go to somewhere which we shall, for the moment, call **Continue66**. Make the necessary changes to the **Adder** section.

Exercise 3.29 Which *nine* lines make up the section after **Adder**? (You will need a loop in these lines, I have called mine **Ran66**).

We have left the most complicated number generator to the end. We must get a number between 1 and 49, however we cannot have the same number more than once in every consecutive group of six. It is therefore necessary to store the previous five lottery numbers to compare the current number with, but let us first get our number between 1 and 49.

Exercise 3.30 Which *five* lines will get a random number between 1 and 49?

When you are doing the same thing to a string of file registers, it is convenient to use indirect addressing. We will therefore make the file registers which store the previous lottery numbers (I have called them **Lott1** to **Lott5**), consecutive file registers. It is easier to test to see whether the number in a file register has reached a certain value by simply testing one bit, rather than going through the business of subtracting the value and testing the zero flag, etc. If **Lott5** is file register number **1F**, going one too far as the PIC scrolls through the five file registers could be tested by looking at bit number 5 in the **FSR**. **Lott1** to **Lott5** should therefore specifically be file registers numbers 1B to 1F. The **FSR** should first be loaded with the number of the first file register (1B). The number in **Random** should then be compared with that in the **Indirect Address** (**INDF**). If the two are the same, the PIC should jump to a section to change the random number (you could call it **Changer**), otherwise it should continue.

Exercise 3.31 Challenge! Which *six* lines will achieve this? You could call this section **CompareLott**.

How can we get a new lottery number? One of the easiest solutions is adding the value in **Skin** to the number in **Random**, and then going through the 1–49 conversion process all over again.

Exercise 3.32 Write the *three* lines which make up **Changer**. (At the end it should go to **RanLot**.) It might be advisable to put **Changer** after the subroutines and before **Start**.

We should now move on to the next lottery value by incrementing the **FSR**. We need to check whether we have used all five registers; this can be done by testing bit 5 of the **FSR**. If we haven't compared the number with all five, the PIC should loop back to **CompareLott**.

Exercise 3.33 Write the *three* lines which will perform this task.

Now that it has been confirmed that the new number is different from any of the others, we need to store it for comparison with later numbers. We will need a file register to keep track of which lottery number we are on (1–6). The device should display this number along with the one between 1 and 49, e.g. **2-37** shows that this is the second number of six, and the number is 37. I've called this register **LottCount**. It is reset in the **Init** subroutine, and then it is incremented after each lottery number (and reset after the sixth). To work out which file register (**Lott1** to **Lott5**) to store the number in, we simply add the hexadecimal number **1B** to the number in **LottCount** (a number between 0 and 5), and move that number into the **FSR**, thus selecting one of **Lott1** to **Lott5**. The random number is then moved into the **INDF** and sent to the appropriate location.

Exercise 3.34 Which *six* lines will store the random number correctly, and then increment **LottCount**.

Now that we have got our random number, we can begin to display it. Our problem is converting a number (e.g. 37) into its tens and ones digits (3 and 7). The tens digit is the number of times we can subtract ten until the result is negative, minus one. The ones digit is number left when we subtract ten enough times. The loop increments a file register holding the tens value, then subtracts the number 10 from **Random**; if the result is negative it stops (skips out of the loop), otherwise it loops back and does it again. We don't want the PIC to increment the tens digit the first time round, so when we jump to this loop from the ends of **Adder** and **RanLot**, we should skip the first instruction. The file register holding the tens value could be called **Tens**, and the loop **TensLoop**.

Exercise 3.35 Write the instruction to jump to **TensLoop** and skip one instruction, at the end of the **Adder** and **RanLot** sections. Then write the first *five* lines of **TensLoop**.

The number in the working register now is 10 less than the units digit of the random number.

Example 3.4 The number was 26, 10 was subtracted three times, and the file register **Tens** was incremented twice. The number in **Tens** is now 2 (the tens digit), and the number left in **Random** is –4 (252 because there are no negative numbers).

We simply need to add 10 to the number in **Random** to get our units digit. In all modes, the units digit is displayed on the display furthest to the right (**dig4**), so we can then convert the units digit into a seven-segment code (using a decoding subroutine) and then move it into **dig4**. We can also clear **dig1, dig2** and **dig3** leaving the three left hand digits blank. It is the line which calls the decoding subroutine which the PIC should jump to after the **Ran66** section, so this line should be labelled **Continue66.**

Exercise 3.36 Write the *seven* lines to do all this. Then write the *eleven* line subroutine which converts the units digit into seven-segment code.

We could now simply use the number in **Tens**, turn it into seven-segment code, and then move it into **dig3**. However, it would look quite nice if we got rid of leading zeroes (displaying _4 instead of 04). Thus we move the number from **Tens** into the working register, and then before calling the **Decoder** subroutine, we should test the zero flag, and skip to the next section (called **LottCounter** if it is set). Otherwise, the PIC should skip that line and call the **Decoder** subroutine. If the device is in any mode except the **-6-6**, the tens digit is displayed on **dig3**, however when in the **-6-6** mode, the tens digit is the other number, and so should be displayed as far away as possible (**dig1**). The mode number for **-6-6** is 4, so we need to see whether or not the number in **Chooser** is 4 (i.e. bit 2 of **Chooser** is set). We should then move the tens digit into **dig3** or **dig1** depending on the result of the test.

Exercise 3.37 Write the *nine* lines to complete these tasks. At the end of the section the PIC can jump to a section called **LottCounter.** Afterwards you may add a section used specifically to move the seven-segment code into **dig3**, which then leads on to **LottCounter.**

LottCounter is a section in which **Chooser** is checked to see whether or not the device is in the lottery mode. If so, the actual lottery number (1 to 6) should also be displayed in the following format: **n-** (e.g. **3-39** means that the third lottery number is 39). Conveniently, when in the lottery mode, **Chooser** holds the number 1, so we can simply apply the instruction **decfsz, w** to **Chooser**, to see if the device is in lottery mode. If it isn't, the PIC can return to **Main** (the numbers have been chosen and displayed), otherwise we will have to put something in **dig1** and **dig2**. Before we return to **Main**, we must reset the file register **Tens**, which is incremented to find the tens digit. This can be done at the beginning of the **LottCounter** section.

Exercise 3.38 What *three* lines will reset the **Tens** file register, and then make the PIC jump back to **Main** if the device is not in lottery mode?

It is now necessary to move the code into **dig2** that will display a dash (turn on the g segment). We can also use the number in **LottCount**, to display the correct number through **dig1**. **LottCount** was 0 before the first digit was chosen, after which it was incremented to 1. We can therefore use the number in **LottCount** directly, turn it into seven-segment code using the **Decoder** subroutine, and then move it into **dig1**.

Exercise 3.39 Write the *five* lines which set up **dig1** and **dig2** with the correct values.

After this the PIC should return to **Main**, clearing the lottery file registers if necessary.

We have (at last) completed the main body of the program, along with various subroutines, but there remain some to be done: **display**, **Chaser** and **MessageChooser**.

The **display** subroutine should use the two least significant bits in **TMR0** to jump to one of four sections. These would turn on a particular display and move the relevant number (from **dig1**-**dig4**) into Port B. After each section the PIC should return, enabling the global interrupt.

Exercise 3.40 What *27* instructions complete the entire **display** subroutine? (**Note:** Put the four sections at the very end of the whole program, because otherwise you may have trouble with the program counter and page boundaries.)

The second subroutine we need to write is the one which causes the segments to chase while the contacts are being pressed. This subroutine is constantly being called during this time, but we want the segments to change every tenth of a second. It therefore needs some sort of timing element, a postscaler (**Mark240**) which allows the PIC to return if a tenth of a second hasn't passed, and continue otherwise. A 2.47 MHz crystal is being used, so a value around 240 would be a suitable postscaler.

Exercise 3.41 Which *six* lines will perform this task (don't forget to update the marker after the correct time has passed)?

We can create the impression of a chase using only three different positions, as illustrated in Figure 3.11.

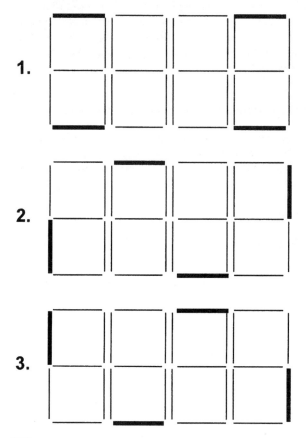

Figure 3.11

Exercise 3.42 Write the codes which need to be in the four displays for each of the three different cases (a total of twelve numbers), and clearly label them.

We will need a file register (**ChaseCount**) to keep track of which version is being displayed (this should be incremented every time a tenth of a second passes), and then used to jump to one of **Chase1, Chase2** or **Chase3**. In **Chase3**, the number 255 should be moved into **ChaseCount** to reset it.

Exercise 3.43 Write the *35* instructions which make the PIC go to one of **Chase1** to **Chase3**, and which make up the three sections in which numbers are moved into file registers **dig1**-**dig4**. At the end of each section, the PIC should return.

Finally, we need to complete the **MessageChooser** subroutine. This uses the number in **Skin** to select a message (maximum of four letters) which is displayed for 3 seconds after the contacts have been released. You can have as many messages as you want, and can adapt the ones suggested in this example.

I have chosen my critical values to be 11–12, 13–15, 16–20, 21–25, 26–35, 36–50 and 50+ (a total of 7). The respective messages will be: sad, bad, cool, john, hot, tops, and ace. We know that the value must be greater than 10, because otherwise the PIC would believe that the contacts aren't pressed, and not call **MessageChooser**.

Exercise 3.44 What *four* lines will jump to a section called **sad** if the number in **Skin** is less than 12 (to be put at the start of the **MessageChooser** subroutine)?

Exercise 3.35 Write the other *21* instructions which will allow the PIC to go to **bad**, **cool**, **john**, **hot**, **tops** or **ace** depending on the value in **Skin**.

Each of the sections **sad** to **ace** must then move the correct numbers in **dig1** to **dig4**. You are best off using a look-up table to find out the codes for the letters. Each letter will have its own subroutine which returns with the appropriate seven-segment code.

Example 3.5 The subroutine _**E** would consist of:

```
_E    retlw    b'10011110'    ; returns with correct seven-segment code
```

Exercise 3.46 Write the letters' look-up table. Include upper and lower case letters and don't forget to start the name of each subroutine with an underscore (_), so that letters such as **b**, and **c** etc., aren't confused by the assembler as the **b**ranch instruction, or the **c**arry flag. My version contained *26* instructions, but yours may differ (don't forget to have a blank).

The sections **sad** to **ace**, now need only call the appropriate subroutine and then move the working register in one of **dig1** to **dig4**.

Exercise 3.47 Write the seven sections **sad** to **ace** which set up **dig1** etc., and then return. They should each consist of *nine* instructions.

All that remains is to define all the general purpose file registers (unless you have been doing so all along), and to set them up correctly in the **Init** subrou-tine. You will find you have to move some sections to the end of your program due to problems with adding to the program counter. For example, if we add the number in **Chooser** to the program counter in the upper half of the page (instruction addresses 100–1FF), we will come across problems. If you want to check your version, the entire program is shown in Program O in Chapter 7.

There are various ways in which you can personalize this device. You could for example change the mode that it first started up in (not a huge difference as the device only 'starts up' when you first connect the batteries). This is done by changing the number you move into **Chooser** in the **Init** subroutine. You could also change the number and content of messages displayed.

4
The new P12C50x series (8 pin PICs)

Why use the P12C50x series?

Early in 1997, a brand new PIC was released – a tiny little 8 pin device offering the same sort of options as the PIC165x series. Some of the surprising features of this little PIC include the fact that it offers up to 6 I/O pins – quite remarkable considering it needs two power supply pins. It needs no external oscillator (e.g. crystal or RC), and even offers a kind of interrupt function. For any application where a small size is needed, and everything can be done using 6 pins or less, the P12C50x series are invaluable.

The series consists of two particular PICs, the P12C508, and P12C509, shown in Figure 4.1. They differ in the amount of program memory they have (the PIC08 can have up to 512 instructions, whereas the PIC09 can hold 1024), and the number of general purpose file registers which they offer.

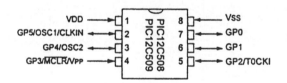

VDD → 1 8 ← Vss
GP5/OSC1/CLKIN ↔ 2 7 ↔ GP0
GP4/OSC2 ↔ 3 6 ↔ GP1
GP3/MCLR/VPP → 4 5 ↔ GP2/T0CKI

PIC12C508
PIC12C509

Figure 4.1

The main differences

There are a few differences in the way the PIC works; these are primarily to do with the file registers. Figure 4.2 shows the file register arrangement for the P12C50x series.

The STATUS register

The first difference is found in the STATUS register. This new PIC offers the option of waking the PIC up if one of three I/O pins changes state (GP0, 1 or 3). The previously unused bit 7 of the STATUS register can now be used to see whether the PIC was woken up from sleep due to one of these pins changing state (bit 7 is *set*), or whether it was for some other reason (bit 7 is *cleared*).

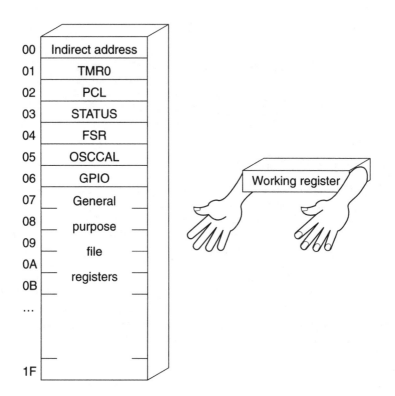

Figure 4.2

The OSCCAL register

Second, you will notice that there is a new file register at address **05**, the **OSCCAL** file register. This is used for **osc**illator **cal**ibration, and is really only used at the start of your program (address **0000**). To make the internal 4 MHz oscillator more accurate, a special number should be moved into the **OSCCAL** register. As with the PIC5x series, the PIC first executes the instruction at the last address (1FF for the PIC508, and 3FF for the PIC509). However, when the PICs are made in the factory, a special instruction is programmed into them at the last address. This instruction moves a particular number (the calibration value) into the working register, i.e. it takes the form:

movlw   xx   ; moves a number xx into the working reg.

After executing this instruction, the PIC loops back and executes the instruction at address 00 (remember, all this happens automatically). If you wish to make the internal oscillator more accurate, the instruction at address 00 should be:

movwf**OSCCAL**; **uses the pre-programmed value to**
; **calibrate the internal oscillator**

Otherwise you can simply start with the usual instruction:

goto**Start**; **goes to beginning of the program**

You therefore no longer need to use the instruction:

org**1FF**

As with the PIC71, you can replace it with:

org**0**

Inputs and outputs

Rather than having I/O ports, the P12C50x series provides us with one file register called the **GPIO** (the general purpose input/output file register). It works exactly the same as the ports, i.e. certain pins on the PIC correspond to bits in this file register.

The OPTION register

As previously mentioned, the PIC can be set up so as to wake up from sleep when one of GP0, GP1 or GP3 change state. This is controlled by bit 7 of the **OPTION** register – the function is *enabled* when bit 7 is *clear*, and *disabled* when the bit is *set*.

Bit 6 of this file register has also been given a purpose (you may remember that these two bits were unused in the PIC5x series). When set, the PIC will make pins GP0, GP1 and GP3 *float* high when not connected to anything, and when the bit is clear, they will not. The rest of **OPTION** is as with the PIC5x series.

The TRIS register

Nothing much is new in this file register; just remember, that there are now 6 bits in the I/O file register, and the number you use to select inputs and outputs should reflect this accordingly.

The general purpose file registers

The P12C508 is almost identical to the PIC54 as far as general purpose file register are concerned (i.e. they go from address 07 to 1F). The P12C509 on the other hand has an extra 16 GPF registers, in a second *bank*. You should remember the principle of an extra bank for file registers from when we looked at the PIC71. File registers 00 to 0F are the same in both banks, but 10 to 1F exist as different file registers in both banks. The bit that chooses which bank you are in is **bit 5** in the **FSR**, set it and the PIC goes into bank 1 mode, and clearing will take it back into bank 0.

The MCLR

The P12C50x series still has an MCLR pin, but if you don't need a reset pin, it can be used as an input pin (**GP3**). You can enable or disable the MCLR when programming the PIC. There is still an input for the TMR0 to count signals from (should you wish to do so), but this can also be made into an I/O pin. This leaves us with two pins devoted to I/O, and four which can be made to be I/O pins.

A template for the P12C50x

```
;***********************************
; written by:                    *
; date:                          *
; version:                       *
; file saved as:                 *
; for PIC...                     *
; clock frequency:               *
;***********************************

;PROGRAM: _____
; _____

        list    P=12C50x
        include "c:\pic\p12c50x.inc"

;============
; Declarations:

        org     0
        (movwf  OSCCAL      ; calibrates internal oscillator)
        goto    Start
```

```
;============
; Subroutines:

Init    clrf     GPIO            ; resets input/output port
        movlw    b'xxxx'         ; sets up which pins are inputs and which
        tris     GPIO            ;   are outputs

        movlw    b'xxxxxxxx'     ; sets up timer and certain pin settings
        option
        retlw    0               ; returns from subroutine


;=============
; Program Start:

Start
        call     Init
Main
        (Write your program here)
END
```

Sample project

Our project to practise using the P12C508 will be a small electronic 'diamond' brooch. It takes advantage of the P12C508's special features, as a brooch should be small, and it uses an interrupt from a tilt switch to make it automatically turn on when picked up, and turn off after being left alone for a certain amount of time.

Adapting the template

We will be using the P12C508, so on the line with the list instruction, replace the x with an 8. The same should be done with the line below (with the include instruction), however remember to adapt the path inside the inverted commas according to the folder in which you have placed the PIC software files.

Inputs and outputs

We will create a sparkling effect using four bi-colour LEDs in a diamond arrangement, constantly and randomly flashing red and green. The circuit diagram in Figure 4.3 shows how everything is arranged. (Remember, bit 3 in GPIO must always be an input.) You can set up the inputs and outputs in the **Init** subroutine using the **tris** instruction. The flowchart would be as shown in Figure 4.4.

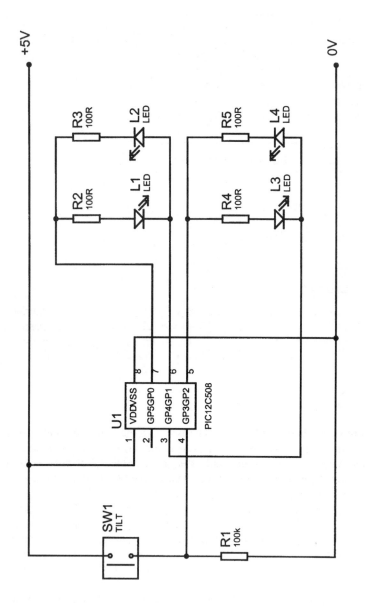

Figure 4.3

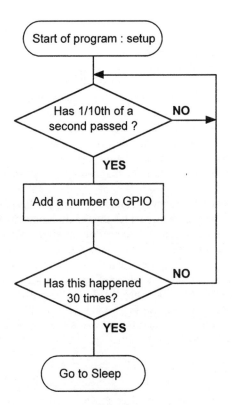

Figure 4.4

Timing

We should now set up TMR0 using the **option** instruction. We will be using TMR0 to create a delay of a tenth of a second, and also to wait 25 seconds, so we should prescale it by the maximum amount. Thinking also about how the PIC is going to be woken up from sleep by the tilt switch, we could also enable the function which wakes the PIC from sleep when one of GP0, 1 or 3 change state. This feature will be ideal, so we should keep bit 7 of the OPTION register clear (remember, clearing the bit enables the function). After setting up the OPTION register, the **Init** subroutine is complete.

Main

In the section labelled Main, the first thing to be done is waiting a tenth of a second. This can be done in a subroutine, which we could call **Timer**. In this subroutine, we need to wait for TMR0 to increment 255 times (i.e. wait for it to overflow). The easiest way to do this is by taking the number out of TMR0 and testing the zero flag in the STATUS register. If the number in TMR0 is zero, the flag will be set, and will otherwise be clear. When it does become zero, you should then increment TMR0 to stop the program reacting more than once to the TMR0 being zero (remember, the TMR0 stays at a particular number for 256 instructions before increasing by one). If the number in TMR0 isn't zero, the PIC should loop back to the start of **Timer**. When a tenth of a second has passed, and the TMR0 has been incremented (by you), the PIC can return from the subroutine. You can now add the decimal number 27 to the general purpose input/output (GPIO) register, thus creating the effect of random flashing. You could add any number you want, and may want to play around a bit to get the best effect.

You should now see whether or not no movement has been detected for about 25 seconds. This is best done by first testing the tilt input, and then skipping the next line if the tilt input is clear. There needs to be some sort of general purpose file register to keep track of how many consecutive times a tenth of a second has passed without a tilt being detected. This could be called **OffTimer**. You should reset it (clear it) in the line after the testing instruction so that it is reset when a tilt is detected. After that, use the **decfsz** instruction to see whether it has gone 256 consecutive tenths of a second with no tilting. Loop back up to **Main** if **OffTimer** hasn't reached 0, and make the PIC go to sleep it if has. After the sleep command don't forget to write:

goto   Main

The program is now complete, and is shown in its entirety in Program P in Chapter 7.

5
A brief introduction to the P16C84

The P16C84 is unique in the way which it is electrically erasable, and thus can be programmed over and over again, without the need to be erased (e.g. by UV light), thus UV erasable PIC84s do not exist. They can therefore save you money, because you don't need to buy a UV eraser (and there is also the fact that the UV erasable versions tend to cost more than the normal ones). The PIC84 also offers a system of memory – EEPROM – which can be read and written to electronically (i.e. by the circuit around it), and thus has four new file registers regarding this function.

The PIC84 is basically a PIC71, but without anything to do with analogue to digital conversion. It therefore has all the file registers to do with interrupts, and the relevant bits (e.g. bit 6 of OPTION) which control them. The do not have an ADRES, ADCON0 or 1, but instead those two spaces are filled with EEDATA and EEADR (in bank 0), and EECON1 and EECON2 (in bank 1). I won't go into the details of these, as the concepts are quite complex and specialized. With your knowledge of how to read the PIC databook's diagrams and the jargon involved, you should nevertheless be able to get any information which you require from the databook. To summarize however, EEDATA holds the actual 8 bit data for the read or write operation, and EEADR holds the address for the EEPROM (a special part of the memory) involved. EECON0 and 1 are control file registers for the EEPROM functions, with each bit corresponding to a particular setting or parameter.

6
Looking to the future

The field of programmable microcontrollers is doubtless one of the fastest growing areas in electronics design, and there is no doubt that new PICs will be coming out all the time, as indeed they have been doing since they first started many years ago. New microcontrollers (apart from Microchip's PIC) are flooding the market, each with their own competitive edge, and all fighting for market domination. The challenge from the user's point of view is keeping up with all these newcomers. As far as new PICs are concerned, they will all have the same structure, but with new additions here and there. A new file register which will perform a special function, extra I/O pins, more timers (such as TMR0) etc. The key to keeping up with these is the data that comes with them. At first sight these enormous manuals seem uncipherable, but there are certain pages to look out for when trying to find out about a new feature. Such pages include ones showing the PIC's file registers, which will allow you to spot any new ones. You can then use the index to find the pages concerning these new file registers which should clear things up.

How to become a better programmer

The key to this is very simple indeed – practice. All that it takes to be able to write programs efficiently and effectively is a bit of experience. Now that you have the knowledge to start writing your own programs you will find that you learn more and more. For example, as I was writing the diamond brooch project I came across something I had never realized before. I wanted to test to see whether or not TMR0 held the number zero, so I wrote the following:

| movf | TMR0 | ; is the number in TMR0 , zero? |
| btfss | STATUS, Z | ; |

I found while simulating the program, that TMR0 just wouldn't count up. As I saw it, the PIC was taking the number out of TMR0, and then putting it back in again. However it then occurred to me that there needs to be something keeping track of exactly what the number in TMR0 is (e.g. 56 and three quarters). It would appear that the integer part of the number (56) is held in TMR0, and the fraction somewhere else. However it also became clear that whenever you move a number into TMR0, that fraction part gets cleared to zero. This explained why TMR0 was never getting anywhere, so I added that all important **w**:

```
movwf      TMR0                 ; is the number in TMR0, zero?
etc.
```

As you can see, you never stop learning – and don't stop experimenting. The great thing about PIC is that you can try things out easily, and then forget it if it didn't work. I recently tried using the PIC to create sounds (though I must admit I began rather pessimistically), but fortunately ended up with some quite effective noises. Be sure to regularly visit the 'LearnPIC' website written by James Cohen and myself. There any of your questions will be answered and you will find other helpful hints, program segments and much more. The wbsite address is: http://members.aol.com/LearnPIC. So with this last piece of advice I leave you, good luck, and happy PICing.

7
Sample programs

Program A

```
;***********************************
; written by: John Morton          *
; date: 21/07/97                   *
; version: 1.0                     *
; file saved as: LedOn             *
; for PIC54                        *
; clock frequency: 3.82 MHz        *
;***********************************

; PROGRAM FUNCTION: To turn on an LED

            list      P=16C54
            include   "c:\pic\p16c5x.inc"

;============
; Declarations:

            porta     equ       05

            org       1FF
            goto      Start
            org       0

;============
; Subroutines:

Init        clrf      porta       ; resets Port A
            movlw     b'0000'     ; RA0: LED, RA1-3: not connected
            tris      porta
            retlw     0
```

```
;=============
; Program Start:

Start
          call      Init            ; sets up inputs and outputs
Main
          bsf       porta, 0        ; turns on LED
          goto      Main            ; loops back to Main

END
```

Program B

```
;************************************
; written by: John Morton            *
; date: 21/07/97                     *
; version: 1.0                       *
; file saved as: PushButtonA         *
; for PIC54                          *
; clock frequency: 3.82 MHz          *
;************************************

; PROGRAM FUNCTION: If a push button is pressed an LED is turned on

          list      P=16C54
          include   "c:\pic\p16c5x.inc"

;============
; Declarations:

          porta     equ       05
          portb     equ       06

          org       1FF
          goto      Start
          org       0

;===========
; Subroutines:

Init      clrf      porta     ; resets inputs and outputs
          clrf      portb     ;
          movlw     b'0000'   ; RA0: LED, RA1-3: not connected
          tris      porta
```

```
          movlw    b'00000001'  ; RB0: push button, RB1-7: not
          tris     portb        ;   connected
          retlw    0
```

;=============
; Program Start:

```
Start     call     Init

Main      btfss    portb, 0     ; tests push button, skip if pressed
          goto     LEDOff       ; push button isn't pressed so turns
                                ;   LED off
          bsf      porta, 0     ; push button is pressed so turns LED on
          goto     Main         ; loops back to Main
LEDOff    bcf      porta, 0     ; turns LED off
          goto     Main         ; loops back to Main
END
```

Program C

```
;***********************************
; written by: John Morton          *
; date: 21/07/97                   *
; version: 2.0                     *
; file saved as: PushButtonB       *
; for PIC54                        *
; clock frequency: 3.82 MHz        *
;***********************************
```

; PROGRAM FUNCTION: If a push button is pressed an LED is turned on

```
          list     P=16C54
          include  "c:\pic\p16c5x.inc"
```

;============
; Declarations:

```
          porta    equ      05
          portb    equ      06

          org      1FF
          goto     Start
          org      0
```

```
;============
; Subroutines:

Init        clrf     porta      ; resets inputs and outputs
            clrf     portb      ;
            movlw    b'0000'    ; RA0: LED, RA1-3: not connected
            tris     porta
            movlw    b'00000001' ; RB0: push button, RB1-7: not
            tris     portb      ;   connected
            retlw    0


;=============
; Program Start:

Start       call     Init

Main        movfw    portb      ; copies the number from Port B into
            movwf    porta      ;   the working register and then back
                                ;   into Port A
            goto     Main       ; loops back to Main

END
```

Program D

```
;***********************************
; written by: John Morton          *
; date: 26/07/97                   *
; version: 1.0                     *
; file saved as: Timing            *
; for PIC54                        *
; clock frequency: 2.4576 MHz      *
;***********************************

; PROGRAM FUNCTION: The state of an LED is toggled every second and
; a buzzer sounds for one second every five seconds.

            list     P=16C54
            include  "c:\pic\p16c5x.inc"
```

```
;============
; Declarations:

                porta    equ        05
                portb    equ        06

                Mark30   equ        08
                Post80   equ        09
                _5Second equ        0A

                org      1FF
                goto     Start
                org      0

;===========
; Subroutines:

Init            clrf     porta      ; resets inputs and outputs
                clrf     portb      ;
                movlw    b'0000'    ; RA0: LED, RA1-3: not connected
                tris     porta
                movlw    b'00000000' ; RB0: buzzer, RB1-7: not connected
                tris     portb

                movlw    b'00000111' ; sets up timing register
                option

                movlw    d'30'      ; sets up marker
                movwf    Mark30     ;

                movlw    d'80'      ; sets up first postscaler
                movwf    Post80     ;

                movlw    d'5'       ; sets up five second counter
                movwf    _5Second   ;

                retlw    0

;=============
; Program Start:

Start           call     Init
```

Main	movfw	Mark30	; takes the number out of Mark30
	subwf	TMR0, w	; subtracts this number from the
			; number in TMR0, leaving the result
			; in the working register (and leaving
			; TMR0 unchanged)
	btfss	STATUS, Z	; tests the zero flag - skip if set, i.e. if the
			; result is zero it will skip the next
			; instruction
	goto	Main	; if the result isn't zero, it loops back to
			; 'Loop'
	movlw	d'30'	; moves the decimal number 30 into the
	addwf	Mark30, f	; working register and then adds it to
			; 'Mark30'
	decfsz	Post80, f	; decrements 'Post80', and skips the
			; next instruction if the result is zero
	goto	Main	; if the result isn't zero, it loops back to
			; 'Loop'
			; one second has now passed
	movlw	d'80'	; resets postscaler
	movwf	Post80	;
	comf	porta, f	; toggles LED state
	bcf	portb, 0	; turns off buzzer
	decfsz	_5Second, f	; has five seconds passed?
	goto	Main	; no, loop back
	bsf	portb, 0	; turns on buzzer
	movlw	d'5'	; resets 5 second counter
	movwf	_5Second	;
	goto	Main	; loops back to start
END			

Program E

```
;**********************************
; written by: John Morton          *
; date: 26/07/97                   *
; version: 1.0                     *
; file saved as: Traffic           *
; for PIC54                        *
; clock frequency: 3.82 MHz        *
;**********************************
```

; PROGRAM FUNCTION: A pedestrian traffic lights junction is simulated.

```
            list      P=16C54
            include   "c:\pic\p16c5x.inc"

;============
; Declarations:

            porta     equ       05
            portb     equ       06

            Mark15    equ       08
            Post80    equ       09
            Counter16           equ     0A
            Counter8 equ        0B

            org       1FF
            goto      Start
            org       0

;============
; Subroutines:

Init        clrf      porta     ; resets inputs and outputs
            clrf      portb     ;
            movlw     b'0001'   ; RA0: push button, RA1-3: not
            tris      porta     ;   connected
            movlw     b'00000000' ; Motorists: RB0: red, RB1: amber,
            tris      portb     ;   RB2: green. Pedestrians: RB4: green,
                                ;   RB5: red, RB3 and RB6-7: not
                                ;   connected
            movlw     b'00000111' ; sets up timing register
            option

            retlw     0
```

delay	movlw	d'15'	; moves the decimal number 15 into
	movwf	Mark15	; the GPF called Mark15
	movlw	d'80'	; moves the decimal number 80 into
	movwf	Post80	; the GPF called Post80
TimeLoop	movfw	Mark15	; takes the number out of Mark15
	subwf	TMR0, w	; subtracts this number from the
			; number in TMR0, leaving the result
			; in the working register (and leaving
			; TMR0 unchanged)
	btfss	STATUS, Z	; tests the zero flag - skip if set, i.e. if the
			; result is zero it will skip the next
			; instruction
	goto	TimeLoop	; if the result isn't zero, it loops back to
			; 'Loop'
	movlw	d'15'	; moves the decimal number 15 into the
	addwf	Mark15, f	; working register and then adds it to
			; 'Mark15'
	decfsz	Post80, f	; decrements 'Post80', and skips the
			; next instruction if the result is zero
	goto	TimeLoop	; if the result isn't zero, it loops back to
			; 'Loop'
	retlw	0	; when half a second has passed, it
			; returns

```
;=============
; Program Start:
```

Start	call	Init	
Main	movlw	b'00010100'	; motorists: green on, others off
	movwf	portb	; pedestrians: red on, others off
ButtonLoop	btfss	porta, 0	; is the pedestrians' button pressed?
	goto	ButtonLoop	; no, so loops back
	bsf	porta, 1	; turns motorists' amber light on
	bcf	porta, 2	; turns motorists' green light off
	call	delay	; waits half a second
	call	delay	; waits half a second
	call	delay	; waits half a second
	call	delay	; waits half a second

```
          movlw    b'00100001'   ; motorists: red on, amber off
          movwf    portb         ; pedestrians: green on, red off

          movlw    d'16'         ; moves the decimal number 16 into the
          movwf    Counter16     ;   general purpose file register called
                                 ;   Counter16
Loop8     call     delay         ; creates half second
          decfsz   Counter16, f  ; does this sixteen times
          goto     Loop8         ; loops back until eight seconds have
                                 ;   passed

          bsf      portb, 1      ; turns motorists' amber light on
          bcf      portb, 0      ; turns motorists' red light off

          movlw    d'8'          ; moves the decimal number 8 into
          movwf    Counter8      ;   Counter8
FlashLoop movlw    b'00100000'   ; toggles the green pedestrian light
          xorwf    portb, f      ;
          call     delay         ; creates half second delay
          decfsz   Counter8, f   ; makes it happen eight times
          goto     FlashLoop     ; loops back
          goto     Main          ; loops back to start

END
```

Program F

```
;***********************************
; written by: John Morton          *
; date: 17/08/97                   *
; version: 1.0                     *
; file saved as: Counter           *
; for PIC54                        *
; clock frequency: 3.82 MHz        *
;***********************************
```

; PROGRAM FUNCTION: To count the number of times a push button is
; pressed, resetting after the sixteenth signal.

```
        list      P=16C54
        include   "c:\pic\p16c5x.inc"
```

;============
; Declarations:

```
        porta    equ    05
        portb    equ    06
        Counter  equ    08
        org      1FF
        goto     Start
        org      0
```

;============
; Subroutines:

```
Init      clrf     porta           ; resets I/O ports
          movlw    b'11111100'     ; moves the code for a 0 into portb
          movwf    portb           ;

          movlw    b'0001'         ; RA0: push button, RA1–3: not
          tris     porta           ;    connected
          movlw    b'00000000'     ; RB0: not connected, RB1–7: 7 seg.
          tris     portb           ;    code
          retlw    0


_7SegDisp
          addwf    PCL             ; skips a certain number of instructions
          retlw    b'11111110'     ; code for 0
          retlw    b'01100000'     ; code for 1
          retlw    b'11011010'     ; code for 2
          retlw    b'11110010'     ; code for 3
          retlw    b'01100110'     ; code for 4
          retlw    b'10110110'     ; code for 5
          retlw    b'10111110'     ; code for 6
          retlw    b'11100000'     ; code for 7
          retlw    b'11111110'     ; code for 8
          retlw    b'11110110'     ; code for 9
```

```
                retlw    b'11101110'  ; code for A
                retlw    b'00111110'  ; code for b
                retlw    b'10011100'  ; code for C
                retlw    b'01111010'  ; code for d
                retlw    b'10011110'  ; code for E
                retlw    b'10001110'  ; code for F
```

;=============
; Program Start:

Start
```
                call     Init         ; sets up inputs and outputs
```
Main
```
                btfss    porta, 0     ; tests push button
                goto     Main         ; if not pressed, loops back

                incf     Counter      ;
                movlw    d'16'        ; has Counter reached 16?
                subwf    Counter, w   ;
                btfsc    STATUS, Z    ;
                clrf     Counter      ; if yes, resets Counter

                movfw    Counter      ; moves Counter into the working reg.
                call     _7SegDisp    ; converts into 7 seg. code
                movwf    portb        ; displays value
                goto     Main         ; loops back to Main

                END
```

Program G

```
;**********************************
; written by: John Morton          *
; date: 19/08/97                   *
; version: 2.0                     *
; file saved as: Counter           *
; for PIC54                        *
; clock frequency: 3.82 MHz        *
;**********************************
```

; PROGRAM FUNCTION: To count the number of times a push button is
; pressed, resetting after the sixteenth signal.

```
            list      P=16C54
            include   "c:\pic\p16c5x.inc"
```

;============
; Declarations:

```
            porta     equ       05
            portb     equ       06
            Counter   equ       08
            org       1FF
            goto      Start
            org       0
```

;============
; Subroutines:

```
Init        clrf      porta       ; resets I/O ports
            movlw     b'11111100' ; moves the code for a 0 into portb
            movwf     portb       ;

            movlw     b'0001'     ; RA0: push button, RA1–3: not
            tris      porta       ;   connected
            movlw     b'00000000' ; RB0: not connected, RB1–7: 7 seg.
                                  ;   code
            tris      portb
            retlw     0


_7SegDisp
            addwf     PCL         ; skips a certain number of instructions
            retlw     b'11111110' ; code for 0
            retlw     b'01100000' ; code for 1
            retlw     b'11011010' ; code for 2
            retlw     b'11110010' ; code for 3
            retlw     b'01100110' ; code for 4
            retlw     b'10110110' ; code for 5
            retlw     b'10111110' ; code for 6
            retlw     b'11100000' ; code for 7
            retlw     b'11111110' ; code for 8
            retlw     b'11110110' ; code for 9
            retlw     b'11101110' ; code for A
```

```
            retlw       b'00111110'  ; code for b
            retlw       b'10011100'  ; code for C
            retlw       b'01111010'  ; code for d
            retlw       b'10011110'  ; code for E
            retlw       b'10001110'  ; code for F
```

```
;=============
; Program Start:

Start
            call        Init          ; sets up inputs and outputs
Main
            btfss       porta, 0      ; tests push button
            goto        Main          ; if not pressed, loops back

            incf        Counter       ;
            movlw       d'16'         ; has Counter reached 16?
            subwf       Counter, w    ;
            btfsc       STATUS, Z     ;
            clrf        Counter       ; if yes, resets Counter

            movfw       Counter       ; moves Counter into the working reg.
            call        _7SegDisp     ; converts into 7 seg. code
            movwf       portb         ; displays value

TestLoop    btfss       porta, 0      ; tests push button
            goto        Main          ; released, so loops back to Main
            goto        TestLoop      ; still pressed, so keeps looping

            END
```

Program H

```
;************************************
; written by: John Morton           *
; date: 19/08/97                    *
; version: 3.0                      *
; file saved as: Counter            *
; for PIC54                         *
; clock frequency: 3.82 MHz         *
;************************************
```

; PROGRAM FUNCTION: To count the number of times a push button is
; pressed, resetting after the sixteenth signal.

```
            list      P=16C54
            include   "c:\pic\p16c5x.inc"

;=============
; Declarations:

            porta     equ       05
            portb     equ       06
            Counter   equ       08

            org       1FF
            goto      Start
            org       0

;============
; Subroutines:

Init        clrf      porta       ; resets I/O ports
            movlw     b'11111100' ; moves the code for a 0 into portb
            movwf     portb       ;

            movlw     b'0001'     ; RA0: push button, RA1–3: not
                                  ;   connected
            tris      porta
            movlw     b'00000000' ; RB0: not connected, RB1–7: 7 seg.
            tris      portb       ;   code

            movlw     b'00000111' ; TMR0 prescaled at 256
            option                ;
            clrf      Counter     ; resets Counter register

            retlw     0


_7SegDisp
            addwf     PCL         ; skips a certain number of instructions
            retlw     b'11111110' ; code for 0
            retlw     b'01100000' ; code for 1
            retlw     b'11011010' ; code for 2
            retlw     b'11110010' ; code for 3
            retlw     b'01100110' ; code for 4
```

```
        retlw    b'10110110'  ; code for 5
        retlw    b'10111110   ; code for 6
        retlw    b'11100000'  ; code for 7
        retlw    b'11111110'  ; code for 8
        retlw    b'11110110'  ; code for 9
        retlw    b'11101110'  ; code for A
        retlw    b'00111110'  ; code for b
        retlw    b'10011100'  ; code for C
        retlw    b'01111010'  ; code for d
        retlw    b'10011110'  ; code for E
        retlw    b'10001110'  ; code for F
```

```
;=============
; Program Start:

Start
        call     Init           ; sets up inputs and outputs
Main
        btfss    portb, 0       ; tests push button
        goto     Main           ; if not pressed, loops back

        incf     Counter, f     ;
        movlw    d'16'          ; has Counter reached 16?
        subwf    Counter, w     ;
        btfsc    STATUS, Z      ;
        clrf     Counter        ; if yes, resets Counter

        movfw    Counter        ; moves Counter into the working reg.
        call     _7SegDisp      ; converts into 7 seg. code
        movwf    portb          ; displays value

TestLoop  btfsc  portb, 0       ; tests push button
        goto     TestLoop       ; still pressed, so keeps looping

        clrf     TMR0           ; resets TMR0
TimeLoop  movlw  d'255'         ; has TMR0 reached 255?
        subwf    TMR0, w        ;
        btfss    STATUS, Z      ;
        goto     TimeLoop       ; if not, keeps looping
        goto     Main           ; 0.07 seconds has passed, so goes to Main

        END
```

Program I

```
;*************************************
; written by: John Morton            *
; date: 31/07/97                     *
; version: 1.0                       *
; file saved as: StopClock           *
; for PIC54                          *
; clock frequency: 2.4576 MHz        *
;*************************************

; PROGRAM FUNCTION: A stop clock displaying the time in tenths of
;    seconds, seconds, tens of seconds and minutes. The minimum time is one
;    second.

            list      P=16C54
            include   "c:\pic\p16c5x.inc"

;============
; Declarations:

            porta     equ       05
            portb     equ       06

            Mark240 equ         08
            TenthSec equ        09
            Seconds   equ       0A
            TenSecond           equ       0B
            Minutes   equ       0C
            Mark80    equ       0D
            Post30    equ       0E
            General   equ       0F

#define     sec       General, 0

            org       1FF
            goto      Start
            org       0

;============
; Subroutines:

Init        clrf      porta     ; resets inputs and outputs
            clrf      portb     ;
```

```
              movlw    b'0000'        ; RA0–3: 7 Segment Display Select
              tris     porta
              movlw    b'00000001'    ; RB0: Multipurpose push button,
              tris     portb          ;   RB1–7: 7 seg. code

              movlw    b'00000111'    ; sets up timing register option

              movlw    d'10'          ; sets up postscaler
              movwf    Post10

              clrf     TenthSec       ; resets GP file registers
              clrf     Seconds        ;
              clrf     TenSecond      ;
              clrf     Minutes        ;
              clrf     General        ;

              retlw    0
;=====
Timer         movfw    Mark240        ; tests to see if TMR0 has passed
              subwf    TMR0, w        ;   through 240 cycles (i.e. 1/10th of a
              btfss    STATUS, Z      ;   second has passed)
              retlw    0              ; 1/10th of a second hasn't passed, so
                                      ;   returns

              movlw    d'240'         ; 1/10th of a second has passed, so adds
              addwf    Mark240, f     ;   the decimal number 240 to Mark240

              incf     TenthSec, f    ; increments the number of tenths of a
                                      ;   second

              movlw    d'10'          ; tests to see whether TenthSec has
              subwf    TenthSec, w    ;   reached 10 (i.e. whether or not one
              btfss    STATUS, Z      ;   second has passed)
              retlw    0              ; 1 second hasn't passed, so returns

              clrf     TenthSec       ; 1 second has passed, so resets TenthSec
              incf     Seconds, f     ;   and increments the number of seconds

              movlw    d'10'          ; tests to see whether Seconds has
              subwf    Seconds, w     ;   reached 10 (i.e. whether or not ten
                                      ;   seconds have passed)
              btfss    STATUS, Z      ;
              retlw    0              ; 10 seconds haven't passed, so returns
```

```
        clrf    Seconds        ; 10 seconds have passed, so resets
        incf    TenSecond, f ; Seconds and increments the number of
                               ;   tens of seconds

        movlw   d'6'           ; tests to see whether TenSecond has
        subwf   TenSecond, w;    reached 6 (i.e. whether or not one
        btfss   STATUS, Z  ;     minute has passed)
        retlw   0              ; 1 minute hasn't passed, so returns

        clrf    TenSecond      ; 1 minute has passed, so resets
        incf    Minutes, f     ; TenSecond and increments the
                               ;   number of minutes

        movlw   d'10'          ; test to see whether Minutes has
        subwf   Minutes, w   ;   reached 10
        btfss   STATUS, Z  ;
        retlw   0              ; 10 minutes haven't passed, so returns

        clrf    Minutes        ; 10 minutes have passed, so resets
        retlw   0              ;   Minutes and returns.
;======
Display  movlw   b'00000011'  ; ignores all but bits 0 and 1 of TMR0
         andwf   TMR0, w      ;   leaving the result in the working
                              ;   register
         addwf   PCL, f       ; adds the result to the program counter
         goto    Display10th  ; displays tenths of a second
         goto    Display1     ; displays seconds
         goto    Display10    ; displays tens of seconds
         goto    DisplayMin   ; displays minutes

Display10th movfw TenthSec    ; takes the number out of TenthSec
         call    _7SegDisp    ; converts the number into 7 seg. code
         movwf   portb        ; displays the value through Port B

         movlw   b'0010'      ; turns on correct display
         movwf   porta        ;
         retlw   0            ; returns

Display1 movfw   Seconds      ; takes the number out of Seconds
         call    _7SegDisp    ; converts the number into 7 seg. code
         movwf   portb        ; displays the value through Port B

         movlw   b'0001'      ; turns on correct display
```

```
                movwf    porta       ;

                retlw    0           ; returns

Display10       movfw    TenSecond   ; takes the number out of TenSecond
                call     _7SegDisp   ; converts the number into 7 seg. code
                movwf    portb       ; displays the value through Port B

                movlw    b'1000'     ; turns on correct display
                movwf    porta       ;

                retlw    0           ; returns

DisplayMin      movfw    Minutes     ; takes the number out of Minutes
                call     _7SegDisp   ; converts the number into 7 seg. code
                movwf    portb       ; displays the value through Port B

                movlw    b'0100'     ; turns on correct display
                movwf    porta       ;

                retlw    0           ; returns

_7SegDisp
                addwf    PCL, f           ; skips a certain number of instructions
                retlw    b'11111110'      ; code for 0
                retlw    b'01100000'      ; code for 1
                retlw    b'11011010'      ; code for 2
                retlw    b'11110010'      ; code for 3
                retlw    b'01100110'      ; code for 4
                retlw    b'10110110'      ; code for 5
                retlw    b'10111110       ; code for 6
                retlw    b'11100000'      ; code for 7
                retlw    b'11111110'      ; code for 8
                retlw    b'11110110'      ; code for 9

OneSec          movfw    Mark80      ; has 1/30th of a second passed?
                subwf    TMR0, w     ;
                btfss    STATUS, A   ;
                retlw    0           ; no, so returns
                movlw    d'80'       ; yes, so resets postscaler
                addwf    Mark80, f   ;
                decfsz   Post30, f   ; has 1 second passed?
                retlw    0           ; no, so returns
                bcf      sec         ; tells rest of program that 1 second has
                                     ;   passed
```

```
            movlw    d'30'         ; resets first postscaler
            movwf    Post30        ;
            retlw    0             ; returns

;=============
; Program Start:

Start       call     Init          ; sets things up

Main        call     Display       ;
            btfss    portb, 0      ; tests for the start button
            goto     Main          ; not pressed so keeps looping

Release     call     Display       ;
            btfsc    portb, 0      ;
            goto     Release       ;

MainLoop    call     Timer         ; timing subroutine
            call     Display       ; display subroutine
            call     OneSec

            btfsc    sec           ; has 1 second passed?
            goto     MainLoop      ; no, so loops back

            btfss    portb, 0      ; has stop button been pressed?
            goto     MainLoop      ; no, so loops back

            bsf      sec           ; resets the sec bit
Release2    call     Display       ; displays the finished time
            btfsc    portb, 0      ; waits for button to be released
            goto     Release2      ;

Debounce    call     display
            call     OneSec        ; has one second passed?
            btfsc    sec           ;
            goto     Debounce      ; no, so loops back

ResetLoop   call     display
            btfss    portb, 0      ; yes, so tests reset button
            goto     ResetLoop     ; it isn't pressed, so loops back

            bsf      sec           ; resets the sec bit
Release3    call     Display       ;
            btfsc    portb, 0      ; waits for button to be released
```

```
              goto      Release3   ;

FinalLoop     call      display
              call      OneSec     ;
              btfsc     sec        ; has 1 second passed?
              goto      FinalLoop  ; no, so loops back

              goto      Main       ; loops back to the beginning
```

END

Program J

```
;************************************
; written by: John Morton            *
; date: 21/07/97                     *
; version: 1.0                       *
; file saved as: LogicGates          *
; for PIC54                          *
; clock frequency: 3.82 MHz          *
;************************************
```

; PROGRAM FUNCTION: To act as the eight different gates.

```
              list      P=16C54
              include   "c:\pic\p16c5x.inc"

;=============
; Declarations:

              porta     equ       05
              portb     equ       06
              STORE     equ       08
              org       1FF
              goto      Start
              org       0

;===========
; Subroutines:

Init          movlw     b'1111'        ; RA0: secondary input, RA1–3: gate
              tris      porta          ;   select bits
              movlw     b'00000001'    ; RB0: primary input, RB4: output,
              tris      portb          ;   RB1–3 and RB5–7: not connected
              retlw     0
```

;==============
; Program Start:

```
Start
            call    Init            ; sets up inputs and outputs

Main        bcf     STATUS, C       ; makes sure carry flag is clear
            rrf     porta, w        ; bumps off bit 0, leaving the result in
                                    ;    the working register
            addwf   PCL, f          ; adds this number to the program
                                    ;    counter
            goto    NOT             ; the code for a NOT gate is received
            goto    BUFFER          ; the code for a BUFFER is received
            goto    AND             ; etc.
            goto    IOR             ;
            goto    XOR             ;
            goto    NAND            ;
            goto    NOR             ;
            goto    XNOR            ;

NOT         swapf   portb, f        ; swaps bit 0 for bit 4 (input for output)
            comf    portb, f        ; inverts the whole of port b (and thus
                                    ;    also bit 4)
            goto    Main            ; loops back to Main

BUFFER      swapf   portb, f        ; swaps the input bit for the output bit.

            goto    Main            ; loops back

AND         movfw   porta           ; takes the number out of Port A
            andwf   portb, w        ; ANDs the number with Port B
common1     movwf   STORE           ; stores result in a GP file register
            swapf   STORE, w        ; swaps nibbles and put result in
                                    ;    working reg.
            movwf   portb           ; returns result to Port B
            goto    Main            ;

IOR         movfw   porta           ; takes the number out of Port A
            iorwf   portb, w        ; ANDs the number with Port B
            goto    common1         ; goes to the section in AND which is
                                    ;    repeated

XOR         movfw   porta           ; takes the number out of Port A
            xorwf   portb, w        ; ANDs the number with Port B
```

```
              goto     common1     ; goes to the section in AND which is
                                    ;   repeated
NAND          movfw    porta       ; takes the number out of Port A
              andwf    portb, w    ; ANDs the number with Port B
common2       movwf    STORE       ; stores result in a GP file register
              swapf    STORE, f    ; swaps nibbles
              comf     STORE, w    ; inverts result and puts result in
                                   ;   working reg.
              movwf    portb       ; returns result to Port B
              goto     Main        ;

NOR           movfw    porta       ; takes the number out of Port A
              iorwf    portb, w    ; ANDs the number with Port B
              goto     common2     ; goes to NOTing section

XNOR          movfw    porta       ; takes the number out of Port A
              xorwf    portb, w    ; XORx the number with Port B
              goto     common2     ; goes to NOTing section

END
```

Program K

```
;************************************
; written by: John Morton           *
; date: 07/08/97                    *
; version: 1.0                      *
; file saved as: Alarm              *
; for PIC54                         *
; clock frequency: 2.4576 MHz       *
;************************************
; PROGRAM FUNCTION: An alarm system which can be set or disabled.

              list     P=16C54
              include  "c:\pic\p16c5x.inc"


;============
; Declarations:

              porta    equ    05
              portb    equ    06

              Post256  equ    08

              org      1FF
```

```
            goto        Start
            org         0
```

;============
; Subroutines:

```
Init        clrf        porta       ; resets inputs and outputs
            clrf        portb       ;
            movlw       b'0011'     ; RA0: Sensor, RA1:Settings switch,
            tris        porta       ;    RA2: not connected, RA3: siren
            movlw       b'00000000' ; RB0: green LED, RB1: red LED
            tris        portb       ;    RB2–7: not connected

            movlw       b'00001111' ; WDT prescaled by 128 (TMR0 not
            option                  ;    prescaled)

            clrf        Post256     ; sets up first postscaler

            retlw       0
```

;=============
; Program Start:

```
Start       call        Init

Main        btfsc       porta, 1    ; tests setting switch
            goto        GreenLed    ; switch is high, so turn on green LED
            bsf         portb, 1    ; switch is low, so turn on red LED

TenthSecond movfw       TMR0        ; is TMR0 at 0?
            btfss       STATUS, Z   ;
            goto        Continue    ;

            incf        TMR0        ;
            decfsz      Post256     ;
            goto        Continue    ;
            clrf        portb       ; it has, so turns off all LEDs

Continue    btfsc       porta, 1    ; tests setting switch
            goto        TenthSecond ; disabled, so doesn't test trigger input

            btfss       porta, 0    ; tests to see whether motion sensor has
                                    ;    been set
            goto        TenthSecond ; not triggered, so loops back
```

```
              bsf       porta, 3      ; turns on siren
EndLoop       clrwdt                  ; resets watchdog timer
              goto      EndLoop       ; constantly loops

GreenLed      bsf       portb, 0      ; turns on green LED
              goto      TenthSecond ; loops back to main body of program.
```

END

Program L

```
;************************************
; written by: John Morton          *
; date: 24/08/97                   *
; version: 1.0                     *
; file saved as: Bike              *
; for PIC54                        *
; clock frequency: 3.82 MHz        *
;************************************

; PROGRAM FUNCTION: A bicycle speedometer and mileometer
;

              list      P=16C54
              include   "c:\pic\p16c5x.inc"


;=============
; Declarations:

              porta     equ       05
              portb     equ       06

              Dist1     equ       09
              Dist10    equ       0B
              Dist100   equ       08

              SP10th    equ       0D
              SP1       equ       0F
              SP10      equ       0C

              Speed10th         equ       10
              Speed1    equ       11
              Speed10   equ       12
```

```
                General    equ        13
                Mark89     equ        14
                tempa      equ        15
                _10        equ        16

#define         mode       portb, 0
#define         counter    porta, 3

#define         debouncer             General, 0


                org        1FFh
                goto       Start
                org        0
```

;============
; Subroutines:

```
Init            movlw      b'0001'    ; yes, so resets Port A
                movwf      porta      ;
                clrf       portb
                movlw      b'1000'    ; RA0-2: controllers for 7 seg.
                tris       porta      ;   display, RA3 - counter
                movlw      b'00000001' ; RB0: select switch, RB1-7 7
                tris       portb      ;   seg. code

                movlw      d'9'       ; resets speed regs.
                movwf      Speed10th  ;
                movwf      Speed1     ;
                movwf      Speed10    ;

                clrf       Dist1
                clrf       Dist10
                clrf       Dist100
                clrf       TMR0       ;
                clrf       SP1
                clrf       SP10th
                clrf       SP10
                retlw      0


Display         movwf      FSR        ; speed, or distance
                decfsz     _10        ; changes display every ten times
                retlw      0          ;   it gets here
```

```
              movlw    d'10'        ;
              movwf    _10          ;

              movlw    b'0111'
              andwf    porta, w
              movwf    tempa
              bcf      STATUS, C
              rrf      tempa        ; selects next display
              btfss    STATUS, C
              goto     CodeSelect
              movlw    b'0100'      ; yes, so resets Port A
              movwf    tempa        ;

CodeSelect    movlw    b'0111'      ; ignores button
              andwf    porta,w      ; uses Port A to select correct
              addwf    FSR, f       ;    file register
              movfw    INDF         ; takes out the correct code
              call     _7SegDisp    ; converts code
              movwf    portb        ; displays number
              movfw    tempa
              movwf    porta
              retlw    0            ; returns

_7SegDisp     addwf    PCL          ; returns with correct code
              retlw    b'01111110'  ; 0
              retlw    b'00001100'  ; 1
              retlw    b'10110110'  ; 2
              retlw    b'10011110'  ; 3
              retlw    b'11001100'  ; 4
              retlw    b'11011010'  ; 5
              retlw    b'11111010'  ; 6
              retlw    b'00001110'  ; 7
              retlw    b'11111110'  ; 8
              retlw    b'11011110'  ; 9
              retlw    b'01110000'  ; L

Debounce      btfsc    debouncer    ; has signal finished?
              goto     NextTest     ; yes, so tests button

              btfss    counter      ; has signal finished?
              bsf      debouncer    ; yes, so sets bit
              retlw    0            ; no, so returns

NextTest      btfss    counter      ; second signal?
```

```
            retlw    0               ; no, so returns

            movfw    Speed10th       ; transfers file regs. so that
            movwf    SP10th          ;    values are displayed
            movfw    Speed1          ;
            movwf    SP1             ;
            movfw    Speed10         ;
            movwf    SP10            ;

            movlw    d'9'            ; resets speed regs.
            movwf    Speed10th       ;
            movwf    Speed1          ;
            movwf    Speed10         ;

            bcf      debouncer
            retlw    0
```

```
;=============
; Program Start:

Start
            call     Init
Main
            btfsc    mode            ; which mode is it in?
            goto     Speed           ; Speed mode
;=========
Distance

            movlw    b'00110100'     ; TMR0 counts external signals
            option                   ;    prescaled by 32

DistLoop    btfsc    mode            ; checks mode
            goto     Speed           ; Speed mode

            movlw    07h
            call     Display         ;

            movlw    d'21'           ; has TMR0 reached 21?
            subwf    TMR0, w         ;
            btfss    STATUS, Z       ;
            goto     DistLoop        ; no, so loops back

            incf     Dist1           ; increments 1 kms
            clrf     TMR0
```

```
            movlw    d'10'          ; has Dist1 reached 10?
            subwf    Dist1, w       ;
            btfss    STATUS, Z      ;
            goto     DistLoop       ; no, so loops back

            incf     Dist10         ; increments 10 kms
            clrf     Dist1

            movlw    d'10'          ; has Dist10 reached 10?
            subwf    Dist10, w      ;
            btfss    STATUS, Z      ;
            goto     DistLoop       ; no, so loops back

            incf     Dist100        ; increments 100 kms
            clrf     Dist10

            movlw    d'10'          ; has Dist100 reached 10?
            subwf    Dist100, w     ;
            btfss    STATUS, Z      ;
            goto     DistLoop       ; no, so loops back
            clrf     Dist100        ; has passed limit, so resets and
            goto     Main           ;   loops back

Speed       movlw    b'00000110'    ; TMR0: internal, prescaled
            option                  ;   at 128
            btfss    counter        ; waits for first signal
            goto     Speed+2        ; keeps looping

BasicTimeLoop
            btfss    mode           ; checks mode
            goto     Distance       ; Speed mode

            movlw    0Bh            ;
            call     Display        ;

            call     Debounce       ;

            movfw    Mark89         ; have 0.0185 seconds passed?
            subwf    TMR0, w        ;
            btfss    STATUS, Z      ;
            goto     BasicTimeLoop        ; no, so loops back

            movlw    d'89'          ; (adds 89 to marker)
            addwf    Mark89         ;
```

```
         decf      Speed10th, f ; yes, so decrements speed by
                                 ;   one tenth of a km per hour
         movlw     d'255'       ; has it passed 0?
         subwf     Speed10th, w  ;
         btfss     STATUS, Z  ;
         goto      BasicTimeLoop        ; no, so loops back

         movlw     d'9'          ; resets 10th unit
         movwf     Speed10th  ;
         decf      Speed1, f  ;
         movlw     d'255'        ; has it passed 0?
         subwf     Speed1, w  ;
         btfss     STATUS, Z  ;
         goto      BasicTimeLoop        ;

         movlw     d'9'          ; resets 1 unit
         movwf     Speed1     ;
         decf      Speed10,f  ;
         movlw     d'255'        ; has it passed 0?
         subwf     Speed10, w  ;
         btfss     STATUS, Z  ;
         goto      BasicTimeLoop        ;

TooSlow  clrf      SP10th        ; displays "SLO" on the displays
         movlw     d'10'
         movwf     SP1
         movlw     d'5'
         movwf     SP10
         movlw     0Bh           ;
         call      Display       ;
         btfss     counter       ; tests for button
         goto      TooSlow       ; no, so keeps looping

         movlw     d'9'          ; resets speed regs.
         movwf     Speed10    ;

         goto      BasicTimeLoop        ;

END
```

Program M

```
;***********************************
; written by: John Morton          *
; date: 24/08/97                   *
; version: 1.0                     *
; file saved as: Quiz              *
; for PIC71                        *
; clock frequency: 2.47 MHz        *
;***********************************
```

; PROGRAM FUNCTION: To detect which of four push buttons has been
; pressed first and turn on the corresponding LED. A buzzer is also turned
;   on for one second.

```
                list    P=16C71
                include "c:\pic\p16c71.inc"

;===========
; Declarations:

                porta   equ     05
                portb   equ     06
                Ten     equ     0C

                org     0
                goto    Start
;===========
; Subroutines:


Init            clrf    porta       ; resets input/output ports
                clrf    portb
                goto    InitContinue; skips address 0004
                goto    isr         ; at address 0004, goes to isr
InitContinue
                bsf     STATUS, 5   ; selects bank 1
                clrf    TRISA       ; RA0: buzzer, RA1-RA4: not connected
                movlw   b'11110000' ; RB0-RB3: LEDs, RB4-RB7: push
                movwf   TRISB       ;   buttons

                movlw   b'10000111' ; sets up timing register (Port B pull-ups
                movwf   OPTION_REG     ;   OFF!)
```

```
            bcf      STATUS, 5   ; goes back to bank 0

            movlw    b'10001000'  ; enables RBChange, and global
            movwf    INTCON       ;   interrupts

            movlw    d'10'        ; moves the decimal number 10 into the
            movwf    Ten          ;   GPF called Ten

            return                ; returns from interrupt

;========
isr         btfss    INTCON, 0   ; tests RBChange flag
            b        Timer       ; TMR0 interrupt occurred
                                 ; RBChange interrupt occurred
            swapf    portb, f    ; turns appropriate LED on
            bsf      porta, 0    ; turns on buzzer
            movlw    b'00100000'  ; enables TMR0 interrupt, disables
            movwf    INTCON       ;   RBChange
            retfie                ; returns, setting the global interrupt
                                 ;   enable

Timer       bcf      INTCON, 2   ; resets TMR0 interrupt flag
            decfsz   Ten, f      ; has this happened 10 times?
            retfie                ; no, so returns, enabling the global int.
                                 ;   bit
            bcf      porta, 0    ; yes, so turns off buzzer
            movlw    b'10001000'  ; global and RBChange enabled
            movwf    INTCON       ;
            sleep                 ; goes into low power consuming mode
            return

;=============
; Program Start:

Start
            call     Init        ; sets everything up
Main
            goto     Main        ; keeps looping until interrupted
END
```

Program N

```
;************************************
; written by: John Morton          *
; date: 25/08/97                   *
; version: 1.0                     *
; file saved as: TempSense         *
; for PIC71                        *
; clock frequency: 2.47 MHz        *
;************************************
; PROGRAM FUNCTION: To detect whether bath water temperature is too
;   cold, OK, or too hot.


                list    P=16C71
                include "c:\pic\p16c71.inc"

;============
; Declarations:


                porta   equ     05
                portb   equ     06

                org     0
                goto    Start


;============
; Subroutines:

Init            clrf    porta           ; resets input/output ports
                clrf    portb
                goto    InitContinue    ; skips address 0004
                goto    isr             ; at address 0004, goes to isr
InitContinue
                bsf     STATUS, 5       ; selects bank 1
                movlw   b'00001'        ; RA0: temperature sensor,
                movwf   TRISA           ;   RA1-4: not connected
                clrf    TRISB           ;   RB0-2: LEDs, RB3-7: not connected

                movlw   b'00000010'     ; sets up ADCON1: RA0 and RA1 as
                movwf   ADCON1          ;   analogue inputs, Vref is the supply
                                        ;   voltage

                movlw   b'00000111'     ; sets up timing register
                movwf   OPTION_REG      ; TMR0 prescaled at 256
```

```
        bcf     STATUS, 5      ; goes back to bank 0

        movlw   b'01000001'    ; sets up A/D register:
        movwf   ADCON0         ; clock: Fosc/8, channel: AN0, converter
                               ;   is on

        movlw   b'11000000'    ; sets up interrupts register:
        movwf   INTCON         ; global and A/D interrupts enabled

        return                 ; returns from the subroutine

isr     bcf     ADCON0, 1      ; resets A/D interrupt flag

        movlw   d'18'          ; compares result with the decimal
        subwf   ADRES, w       ;   number 18 without affecting ADRES
        btfss   STATUS, C      ;
        goto    Cold           ; less than 18, so too cold

        movlw   d'23'          ; compares result with the decimal
        subwf   ADRES, w       ;   number 23 without affecting ADRES
        btfss   STATUS, C      ;
        goto    Ok             ; less than 23, so OK
        goto    Hot            ; more than 23, so hot
Cold    movlw   b'00000001'    ; turns on cold LED (others off)
        movwf   portb          ;
        retfie                 ; returns, enabling global interrupt

Ok      movlw   b'00000010'    ; turns o.k. LED (others off)
        movwf   portb          ;
        retfie                 ; returns, enabling global interrupt

Hot     movlw   b'00000100'    ; turns on hot LED (others off)
        movwf   portb          ;
        retfie                 ; returns, enabling global interrupt

;=============
; Program Start:

Start
        call    Init           ; sets everything up
Main    bsf     ADCON0, 2      ; starts A/D conversion
        goto    Main           ; keeps looping until interrupted

END
```

Program O

```
;*************************************
; written by: John Morton          *
; date: 25/08/97                   *
; version: 1.0                     *
; file saved as: RanLott           *
; for PIC71                        *
; clock frequency: 2.47 MHz        *
;*************************************
; PROGRAM FUNCTION: A multipurpose random number generator with
;   various modes including a lottery number function.


                list      P=16C71
                include   "c:\pic\p16c71.inc"


;=============
; Declarations:


                porta       equ       05
                portb       equ       06

                dig1        equ       0C
                dig2        equ       0D
                dig3        equ       0E
                dig4        equ       0F

                Chooser     equ       10
                ChaseCount  equ       11
                Skin        equ       12
                Random      equ       13
                Mark240     equ       14
                LetGoCount  equ       15
                TouchTime   equ       16
                General     equ       17
                Scaler      equ       18
                LottCount   equ       19
                Tens        equ       1A
                ADCount     equ       20
                Post28      equ       21
                Post70      equ       22

                Lott1       equ       1B
                Lott2       equ       1C
```

```
            Lott3        equ        1D
            Lott4        equ        1E
            Lott5        equ        1F

#define     sec3         General, 0

            org          0
            goto         Start
```

```
;===========
; Subroutines:
```

```
Init        clrf         porta        ; resets I/O ports
            clrf         portb
            b            Init+4       ; leaves address 0004 for isr
            b            isr

            movlw        23           ; clears file register from number 23 to
            movwf        FSR          ;    the FSR
ClearLoop   decf         FSR          ;
            clrf         INDF         ;
            movfw        FSR          ;
            btfsc        STATUS, Z    ;
            goto         ClearLoop    ;

            bsf          STATUS,5     ; selects bank 1

            movlw        b'00001'     ; RA0: contacts, RA1-RA4: seven
            movwf        TRISA        ;    segment controlling pins
            movlw        b'00000001'  ; RB0: mode button, RB1-RB7:
            movwf        TRISB        ;    seven segment code

            movlw        b'00000111'  ; sets up TMR0, prescaled at 256
            movwf        OPTION_REG   ;

            movlw        b'00000010'  ; RA0 and RA1 are analogue, Vref is
            movwf        ADCON1       ;    supply voltage

            bcf          STATUS,5     ; returns to bank 0

            movlw        b'00010000'  ; sets up interrupts:
            movwf        INTCON       ; external on, global off

            movlw        b'01000001'  ; sets up A/D conversion:
```

```
                movwf   ADCON0        ; clock: Fosc/8, channel: AN0, converter
                                      ;   is on

                movlw   d'4'          ; so that device starts up in 1-6 mode
                movwf   Chooser       ;

                movlw   d'28'         ; sets up postscalers
                movwf   Post28        ;
                movlw   d'70'         ;
                movwf   Post70        ;

                return                ; returns from subroutine


isr             btfss   INTCON, 1     ; has external interrupt occurred?
                b       TMRInt        ; no, so goes to timer interrupt
                                      ; yes, so continues

                incf    Chooser       ; goes on to next mode
                movlw   d'5'          ; has it gone through all five modes?
                subwf   Chooser,w     ;
                btfsc   STATUS,Z      ;
                clrf    Chooser       ; yes, so reset back to mode 1 (1-6)
                movfw   Chooser       ; takes the number out of Chooser
                addwf   PCL, f        ; skips that many instructions
                b       Dis6          ; 1-6_
                b       DisLot        ; Lott
                b       Dis12         ; 1-12
                b       Dis99         ; 1-99
                b       Dis66         ; -6-6

Dis6            movlw   b'10111110'   ; 6
                movwf   dig3
                movlw   b'00000000'   ; blank
                movwf   dig4
                b       DisCommon     ; goes to common place to set up 1 -

Dis12           movlw   b'11011010'   ; 2
                movwf   dig4
                movlw   b'01100000'   ; 1
                movwf   dig3
                b       DisCommon     ; 1-

Dis99           movlw   b'11100110'   ; 9
```

```
               movwf   dig3
               movlw   b'11100110'  ; 9
               movwf   dig4
               b       DisCommon ; 1-

DisCommon      movlw   b'01100000'  ; 1
               movwf   dig1
               movlw   b'00000010'  ; -
               movwf   dig2
               return               ; returns, without re-enabling the
                                    ;   interrupts

DisLot         movlw   b'00011100'  ; L
               movwf   dig1
               movlw   b'00111010'  ; o
               movwf   dig2
               movlw   b'00001110'  ; t
               movwf   dig3
               movlw   b'00001110'  ; t
               movwf   dig4
               return

Dis66          movlw   b'00000010'  ; sets a dash
               movwf   dig1
               movwf   dig3
               movlw   b'10111110'  ; sets a number 6
               movwf   dig2
               movwf   dig4
               return

TMRInt         bsf     INTCON, 4  ; enables external interrupt

               decfsz  Post28     ; first postscaled by 28
               retfie             ;

               movlw   d'28'      ; yes, so resets first postscaler
               movwf   Post28     ;
               bcf     sec3

               decfsz  Post70     ; secondly postscaled by 70, (3.5 minutes
                                  ;   passed?)
               retfie             ; no

               movlw   d'70'      ; yes, so resets second postscaler
```

```
                movwf   Post70      ;

                clrf    portb       ; clears outputs
                clrf    porta       ;
                sleep               ; sleeps (low power mode)
                retfie              ; returns, enabling the global interrupt
```

;==

```
display         movfw   TMR0        ; use TMR0 to select one of four sections
                andlw   b'00000011' ;
                addwf   PCL         ;
                b       digit1
                b       digit2
                b       digit3
                b       digit4

digit1          movlw   b'10000'    ; turns on relevant display
                movwf   porta       ;
                movfw   dig1
                movwf   portb       ; moves relevant code into port b
                retfie

digit2          movlw   b'00010'    ; turns on relevant display
                movwf   porta       ;
                movfw   dig2
                movwf   portb       ; moves relevant code into port b
                retfie

digit3          movlw   b'00100'    ; turns on relevant display
                movwf   porta       ;
                movfw   dig3
                movwf   portb       ; moves relevant code into port b

                retfie

digit4          movlw   b'01000'    ; turns on relevant display
                movwf   porta       ;
                movfw   dig4
                movwf   portb       ; moves relevant code into port b
                retfie
```

;==

```
Chaser          movfw   Mark240     ; has a tenth of a second passed?
                subwf   TMR0, w     ;
```

```
        btfss    STATUS, Z  ;
        return              ; no, so returns

        movlw    d'240'     ; yes, so continues and resets marker
        addwf    Mark240    ;
        incf     ChaseCount ; scrolls through the 3 different display
        movfw    ChaseCount ;   codes, creating a chase
        addwf    PCL
        b        Chase1
        b        Chase2
        b        Chase3
```

;===

MessageChooser

```
        movlw    d'12'      ; is Skin between 11 and 12?
        subwf    Skin,w     ;
        btfss    STATUS, C  ;
        b        sad        ; yes, so displays SAd
        movlw    d'15'      ; is Skin between 13 and 15?
        subwf    Skin,w     ;
        btfss    STATUS, C  ;
        b        bad        ; yes, so displays bAd

        movlw    d'20'      ; is Skin between 16 and 20?
        subwf    Skin,w     ;
        btfss    STATUS, C  ;
        b        cool       ; yes, so displays cool

        movlw    d'25'      ; is Skin between 21 and 25?
        subwf    Skin,w     ;
        btfss    STATUS, C  ;
        b        john       ; yes, so displays John

        movlw    d'35'      ; is Skin between 26 and 35?
        subwf    Skin,w     ;
        btfss    STATUS, C  ;
        b        hot        ; yes, so displays hot

        movlw    d'50'      ; is Skin between 36 and 50?
        subwf    Skin,w     ;
        btfss    STATUS, C  ;
        b        tops       ; yes, so displays toPS
        b        ace        ; no, so displays ACE (above 50)
```

```
;=================================================================
Decoder     addwf   PCL          ; converts number into 7 seg. code
            retlw   b'11111100'  ; number 0
            retlw   b'01100000'  ; number 1
            retlw   b'11011010'  ; number 2
            retlw   b'11110010'  ; etc.
            retlw   b'01100110'  ;
            retlw   b'10110110'  ;
            retlw   b'10111110'
            retlw   b'11100000'  ;
            retlw   b'11111110'  ;
            retlw   b'11100110'  ;


;=================================================================
Changer     movfw   Skin         ; gets new random number
            addwf   Random       ;
            b       RanLot       ; converts it to a number between 1 and
                                 ;   49
;=================================================================
; Program Start:

Start       call    Init         ; set everything up
            bsf     INTCON, 1    ; tell the isr to go to mode changing
                                 ;   section
            call    isr          ; call isr

Main        bsf     ADCON0,2     ; starts A/D conversion
            call    display      ; sorts out displays
            btfsc   portb, 0     ; tests for up button
            b       ADTest       ; still pressed, so changes nothing
            movlw   b'10100000'  ; released, so enables TMR0 and global
            movwf   INTCON       ;   interrupt

ADTest      btfsc   ADCON0, 2    ; conversion is in progress
            b       Main+1       ; goes back to main, and skips one line

            movlw   d'10'        ; is result less than 10? (are contacts
            subwf   ADRES, w     ;   pressed?)
            btfss   STATUS, C    ;
            b       Main         ; no, so loops back

            movlw   d'28'        ; resets timing registers
            movwf   Post28       ;
            movlw   d'70'        ;
            movwf   Post70       ;
```

```
              bcf      INTCON, 7   ; no more interrupts from now on

Stabilizer
              bsf      ADCON0, 2   ; starts conversion
              btfsc    ADCON0, 2   ; has it finished?
              b        Stabilizer  ; no, so keeps looping

              decfsz   ADCount     ; has this happened 256 times
              b        Stabilizer  ; no, so loops back
              movfw    ADRES       ; yes, so stores A/D result
              movwf    Skin

Main2         bsf      ADCON0,2    ; starts A/D conversion (are contacts
                                   ;    released?)
              call     Chaser      ; sorts out chasing while contacts are
                                   ;    pressed
              call     display     ;
              btfsc    ADCON0,2    ; is conversion finished?
              b        Main2+1     ; no, so loops back
              movlw    d'10'       ; is A/D low enough to suggest finger
              subwf    ADRES,w     ;    has been removed?
              btfsc    STATUS,C    ;
              b        Main2       ; no, so loops back an starts another
                                   ;    conversion

              decfsz   LetGoCount  ; yes, confirms low result 200 times
              b        Main2       ;
              movlw    d'200'      ; resets the counting GPF register
              movwf    LetGoCount  ;

              movfw    TMR0        ; takes the number out of TMR0
              movwf    TouchTime   ; stores it in TouchTime

              call     MessageChooser   ; choose message using number
                                   ;    in Skin
              movlw    d'28'       ; resets first postscaler, so that the PIC
              movwf    Post28      ;    times the full 3 seconds
              bsf      sec3        ;

Loop3Sec      movlw    b'10100000' ; enables TMR interrupt
              movwf    INTCON      ;
              call     display     ; keeps displays going
              btfsc    sec3        ;
```

```
            b        Loop3Sec     ; keeps looping until 3 seconds have
                                  ;   passed

            movfw    TouchTime    ; adds TouchTime and Skin together,
            addwf    Skin, w      ;   storing the result in the GPF, random
            movwf    Random       ;

            movfw    Chooser      ; use the mode to jump to the correct
            addwf    PCL          ;   section
            b        Ran6
            b        RanLot
            b        Ran12
            b        Ran99
            b        Ran6         ; to begin with, it is the same as 1-6 mode
Ran6        movlw    d'6'
            b        Adder
Ran12       movlw    d'12'
            b        Adder
Ran99       movlw    d'99'
            b        Adder

Adder       movwf    Scaler       ; stores the number from Ran6, Ran12,
                                  ;   or Ran99
            movfw    Scaler       ; repeatedly adds the number until it
            addwf    Random       ;   passes 255
            btfss    STATUS,C     ; is the value suitable?
            b        Adder+1      ; no, so loops back
            incf     Random       ; adds one to Random
            btfss    Chooser, 2   ; -6-6 mode?
            b        TensLoop+1   ; no, so carries on

            movwf    Tens         ; stores as first 1-6 value
            addwf    TouchTime    ; gets new random number
            movwf    Random       ;
Ran66       movlw    d'6'         ; converts it into 1-6
            addwf    Random       ;
            btfss    STATUS, C    ;
            b        Ran66        ;
            incf     Random       ;
            b        Continue66   ; jumps to correct place in TensLoop

RanLot      movlw    d'49'        ; repeatedly adds 49 until it goes past
            addwf    Random       ;   255
            btfss    STATUS,C     ; is number suitable?
```

```
              b         RanLot        ; no, so loops back
              incf      Random        ; adds on, so it is between 1 and 49

CompareLott
              movlw     1Bh           ; selects Lott1 first
              movwf     FSR
              movfw     INDF          ; compares with previous lottery values
              subwf     Random, w     ;
              btfsc     STATUS,Z      ;
              b         Changer       ; the two are the same, so gets new
                                      ;   number

              incf      FSR           ; moves on to next lottery number
              btfss     FSR, 5        ; has it gone too far?
              b         CompareLott   ; no, so loops back
                                      ; yes, so continues

              movfw     LottCount     ; use number from LottCount, and add
              addlw     1B            ;   1B to it in order to select the correct
                                      ;   GPF.
              movwf     FSR           ;
              movfw     Random        ;
              movwf     INDF          ;
              incf      LottCount     ; moves on to next lottery number

TensLoop      incf      Tens          ; works the tens digit of the number
              movlw     d'10'         ; (keeps subtracting 10 until result is
              subwf     Random        ;   negative)
              btfsc     STATUS,C      ;
              b         TensLoop      ;

              movlw     d'10'         ; gets units value by adding 10 to
              addwf     Random, w     ;   Random and leaving the result in the
                                      ;   working register.
Continue66    call      Decoder       ; converts units value into 7 seg. code
              movwf     dig4          ; moves code into appropriate digit
              clrf      dig3          ; blanks out left three digits
              clrf      dig2          ;
              clrf      dig1          ;

              movfw     Tens
              btfsc     STATUS, Z     ; checks to see if tens is zero
              b         LottCounter   ; yes, so displays nothing on dig3
```

```
            call     Decoder       ; converts result into 7 seg code
            btfss    Chooser, 2    ; is the device in -6-6 mode?
            b        Not66         ; goes to a place where the tens is shown
                                   ;   on dig3
            movwf    dig1          ; in -6-6 mode, so moves code into dig1
            b        LottCounter   ; now goes to label the lottery number
                                   ;   correctly

Not66       movwf    dig3          ; moves code into dig 3

LottCounter clrf     Tens          ; resets Tens register
            decfsz   Chooser, w    ; is the device in lottery mode?
            b        Main          ; no, so loops back to main
                                   ; yes, so continues
            movlw    b'00000010'   ; displays a dash on digit 2
            movwf    dig2          ;
            movfw    LottCount     ; takes the number out of LottCount
            call     Decoder       ;   and uses it in dig1
            movwf    dig1          ;

            movlw    d'6'          ; has LottCount gone too far?
            subwf    LottCount, w  ;
            btfss    STATUS, Z     ;
            goto     Main          ;
            clrf     LottCount     ; yes, so resets it
            clrf     Lott1         ;
            clrf     Lott2         ;
            clrf     Lott3         ;
            clrf     Lott4         ;
            clrf     Lott5         ;
            goto     Main          ; loops back to the beginning

;====================================================================
sad         call     _S
            movwf    dig1
            call     _A
            movwf    dig2
            call     _d
            movwf    dig3
            call     blank
            movwf    dig4
            return
```

```
bad       call    _b
          movwf   dig1
          call    _A
          movwf   dig2
          call    _d
          movwf   dig3
          call    blank
          movwf   dig4
          return

cool      call    _C
          movwf   dig1
          call    _O
          movwf   dig2
          call    _O
          movwf   dig3
          call    _L
          movwf   dig4
          return

john      call    _j
          movwf   dig1
          call    _o
          movwf   dig2
          call    _h
          movwf   dig3
          call    _n
          movwf   dig4
          return

hot       call    _h
          movwf   dig1
          call    _o
          movwf   dig2
          call    _t
          movwf   dig3
          call    blank
          movwf   dig4
          return

tops      call    _t
          movwf   dig1
          call    _o
```

```
                movwf   dig2
                call    _P
                movwf   dig3
                call    _S
                movwf   dig4
                return

ace             call    _A
                movwf   dig1
                call    _C
                movwf   dig2
                call    _E
                movwf   dig3
                call    blank
                movwf   dig4
                return

_A              retlw   b'11101110'  ; letter A
_b              retlw   b'00111110'  ; letter b
_C              retlw   b'10011100'  ; letter C
_c              retlw   b'00011010'  ; letter c
_d              retlw   b'01111010'  ; letter d
_E              retlw   b'10011110'  ; E
_F              retlw   b'10001110'  ; F
_g              retlw   b'11110110'  ; g
_H              retlw   b'01101110'  ; H
_h              retlw   b'00101110'  ; h
_I              retlw   b'00001100'  ; I
_i              retlw   b'00001000'  ; i
_j              retlw   b'01110000'  ; j
_L              retlw   b'00011100'  ; L
_n              retlw   b'00101010'  ; n
_O              retlw   b'11111100'  ; O
_o              retlw   b'00111010'  ; o
_P              retlw   b'11001110'  ; P
_q              retlw   b'11100110'  ; q
_r              retlw   b'00001010'  ; r
_S              retlw   b'10110110'  ; S
_t              retlw   b'00001110'  ; t
_U              retlw   b'01111100'  ; U
_u              retlw   b'00111000'  ; u
_y              retlw   b'01110110'  ; y
blank           retlw   b'00000000'  ; blank
```

```
;══════════════════════════════════════════════════════════
Chase1      movlw    b'00000100'  ; sets up first chase pattern
            movwf    dig1         ;
            movlw    b'00010000'
            movwf    dig2
            movlw    b'10000000'
            movwf    dig3
            movlw    b'00100000'
            movwf    dig4
            return

Chase2      movlw    b'10010000'  ; sets up second chase pattern
            movwf    dig1         ;
            movlw    b'00000000'
            movwf    dig2
            movlw    b'00000000'
            movwf    dig3
            movlw    b'10010000'
            movwf    dig4
            return

Chase3      movlw    b'00001000'  ; sets up third chase pattern
            movwf    dig1         ;
            movlw    b'10000000'
            movwf    dig2
            movlw    b'00010000'
            movwf    dig3
            movlw    b'01000000'
            movwf    dig4

            movlw    d'255'        ; resets ChaseCount GPF register
            movwf    ChaseCount ;
            return

END
```

Program P

```
;**********************************
; written by: John Morton            *
; date: 21/09/97                     *
; version: 1.0                       *
; file saved as: Diamond             *
; for PIC508                         *
; clock frequency: (internal )  = 4 MHz  *
;**********************************

; PROGRAM FUNCTION: To act like diamond brooch by randomly
;   flashing LEDs

              list      P=12C50x
              include   "c:\pic\p12c50x.inc"

;============
; Declarations:

              OffTimer equ        08

#define       Tilt      GPIO, 3

              org       0
              movwf     OSCCAL    ; calibrates internal oscillator
              goto      Start

;===========
; Subroutines:

Init          clrf      GPIO      ; resets general port
              movlw     b'001000' ; GP0-2, 4: LEDs, GP3: tilt
              tris      GPIO      ; GP5: not connected
              movlw     b'00000111' ; sets up TMR0
              option

              retlw     0

Timing        movfw     TMR0      ; has 1/10th of a second passed?
              btfss     STATUS, Z ;
              goto      Timing    ; no, so keeps looping
              incf      TMR0      ; yes, stops multiple zero read
```

```
        retlw    0              ; returns

;=============
; Program Start:

Start
        call     Init           ; sets up inputs and outputs
Main    call     Timing         ; waits for 1/10th of a second
        movlw    d'27'          ; changes LEDs
        addwf    GPIO           ;
        btfsc    Tilt           ; has tilt signal been received?
        clrf     OffTimer       ; yes, so resets Off Timer
        decfsz   OffTimer       ; no
        goto     Main           ;
        clrf     GPIO           ; turns all LEDs off
        sleep                   ;
        goto     Main           ; loops back up to Main

END
```

Appendix A
Specifications of some PICs

Device	Pins	I/O	ROM	RAM	Max. freq.	Voltage	Features
P12C508	8	6 max	512	25	20 MHz	2.5–6.25	TMR0, WDT, SLEEP, code protection, POR, 2 level stack, interrupts, internal oscillator
P12C509	8	6 max	512	25	20 MHz	2.5–6.25	As P12C508
P16C54	18	12	512	25	20 MHz	2.5–6.25	TMR0, WDT, SLEEP, code protection, POR, 2 level stack
P16C55	28	20	512	24	20 MHz	2.5–6.25	As P16C54
P16C56	18	12	1K	25	20 MHz	2.5–6.25	As P16C54
P16C57	28	20	2K	72	20 MHz	2.5–6.25	As P16C54
P16C71	18	13	1K	36	20 MHz	3–6	4 A/D channels, 8 level stack, interrupts, + those of P16C54
P16C73	28	22	4K	192	20 MHz	3–6	5 A/D channels, TMR1, TMR2, serial interface + those of P16C71
P16C84	18	13	1K	36	10 MHz	2–6	1K EEPROM, 8 level stack, interrupts, + those of P16C54

Appendix B
Pin layouts of common PICs

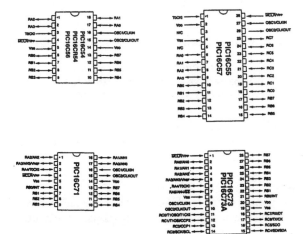

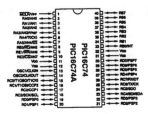

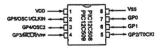

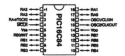

Appendix C
Quick reference

A list of all instructions and what you do them to

addlw   **number**

- (Not for PIC16C5x series) – **add**s a **number** with the number in the working register.

addwf   **FileReg, f**

- **add**s the number in the working register to the number in a file register and puts the result in the file register.

addwf   **FileReg, w**

- **add**s the number in the working register to the number in a file register and puts the result back into the working register, leaving the file register unchanged.

andlw   **number**

- **AND**s a **number** with the number in the working register, leaving the result in the working register.

andwf   **FileReg, f**

- **AND**s the number in the working register with the number in a file register and puts the result in the file register.

bcf   **FileReg, bit**

- clears a bit in a file register (i.e. makes the bit 0).

bsf   **FileReg, bit**

- sets a bit in a file register (i.e. makes the bit 1).

btfsc   **FileReg, bit**

- tests a bit in a file register and skips the next instruction if the result is clear (i.e. if that bit is 0).

btfss   **FileReg, bit**

- tests a bit in a file register and skips the next instruction if the result is set (i.e. if that bit is 1).

call   **AnySub**
- makes the chip **call** a subroutine, after which it will return to where it left off.

clrf   **FileReg**
- **clears** (makes 0) the number in a file register.

clrw
- **clears** the number in the working register.

clrwdt
- **clears** the number in the watchdog timer.

comf   **FileReg, f**
- **com**plements (inverts, ones become zeroes, zeroes become ones) the number in a file register, leaving the result in the file register.

decf   **FileReg, f**
- **dec**rements (subtracts one from) a file register and puts the result in the file register.

decfsz   **FileReg, f**
- **dec**rements a file register and if the result is zero it skips the next instruction. The result is put in the file register.

goto   **Anywhere**
- makes the chip **go to** somewhere in the program which YOU have labelled 'Anywhere'.

incf   **FileReg, f**
- **inc**rements (adds one to) a file register and puts the result in the file register.

incfsz   **FileReg, f**
- **inc**rements a file register and if the result is zero it skips the next instruction. The result is put in the file register.

iorlw   **number**
- inclusive **OR**s a **number** with the number in the working register.

iorwf   **FileReg, f**
- inclusive **OR**s the number in the working register with the number in a file register and puts the result in the file register.

movfw   **FileReg**
or   **movf**   **FileReg, w**
- **mov**es (copies) the number in a file register in to the working register.

movlw   number

- **mov**es (copies) a **number** into the working register.

movwf   FileReg

- **mov**es (copies) the number in a file register into the working register.

nop

- this stands for: **n**o **op**eration, in other words – do nothing (it seems useless, but it's actually quite useful!)

option

- (Not to be used except in PIC16C5x series) – takes the number in the working register and moves it into the **option** register.

retfie

- (Not for PIC16C5x series) – **ret**urns **f**rom a subroutine and enables the Global Interrupt Enable bit.

retlw   number

- **ret**urns from a subroutine with a particular **number** (literal) in the working register.

return

- (Not for PIC16C5x series) – **return**s from a subroutine.

rlf   FileReg, f

- **r**otates the bits in a file register to the **l**eft, putting the result in the file register.

rrf   FileReg, f

- **r**otates the bits in a file register to the **r**ight, putting the result in the file register.

sleep

- sends the PIC to **sleep**, a lower power consumption mode.

sublw   number

- (Not for PIC16C5x series) – **sub**tracts the number in the working register from a **number**.

subwf   FileReg, f

- **sub**tracts the number in the working register from the number in a file register and puts the result in the file register.

swapf   **FileReg, f**
- **swap**s the two halves of the 8 bit binary number in a file register, leaving the result in the file register.

tris   **PORTX**
- (Not to be used except in PIC16C5x series) – uses the number in the working register to specify which bits of a port are inputs (correspond to a binary 1) and which are outputs (correspond to 0).

xorlw   **number**
- exclusive **OR**s a **number** with the number in the working register.

xorwf   **FileReg, f**
- exclusive **OR**s the number in the working register with the number in a file register and puts the result in the file register.

Number system conversion

	0	1	2	3	4	5	6	7	8	9	A	B	C	D	E	F
0	0	16	32	48	64	80	96	112	128	144	160	176	192	208	224	240
1	1	17	33	49	65	81	97	113	129	145	161	177	193	209	225	241
2	2	18	34	50	66	82	98	114	130	146	162	178	194	210	226	242
3	3	19	35	51	67	83	99	115	131	147	163	179	195	211	227	243
4	4	20	36	52	68	84	100	116	132	148	164	180	196	212	228	244
5	5	21	37	53	69	85	101	117	133	149	165	181	197	213	229	245
6	6	22	38	54	70	86	102	118	134	150	166	182	198	214	230	246
7	7	23	39	55	71	87	103	119	135	151	167	183	199	215	231	247
8	8	24	40	56	72	88	104	120	136	152	168	184	200	216	232	248
9	9	25	41	57	73	89	105	121	137	153	169	185	201	217	233	249
A	10	26	42	58	74	90	106	122	138	154	170	186	202	218	234	250
B	11	27	43	59	75	91	107	123	139	155	171	187	203	219	235	251
C	12	28	44	60	76	92	108	124	140	156	172	188	204	220	236	252
D	13	29	45	61	77	93	109	125	141	157	173	189	205	221	237	253
E	14	30	46	62	78	94	110	126	142	158	174	190	206	222	238	254
F	15	31	47	63	79	95	111	127	143	159	175	191	207	223	239	255

Bit assignments of various file registers

OPTION

Bit no.	7	6	5	4	3	2	1	0	
	-	-	RTS	RTE	PSA	PS2	PS1	PS0	Rate
						0	0	0	1:2
						0	0	1	1:4
						0	1	0	1:8
						0	1	1	1:16
						1	0	0	1:32
						1	0	1	1:64
						1	1	0	1:128
						1	1	1	1:256

Prescaler assignment
0 - if you want the prescaler to be used by the TMR0
1 - if you want prescaler to be used by the WDT

TMR0 signal increment
0 - if you want to count up when the signal drops
1 - if you want to count up when the signal rises

TMR0 signal source
0 - if you want to count an internal signal
1 - if you want to count an external signal

(P16C5x) **Unassigned**

(P16C71) **Interrupt edge select**
0 - External interrupt occurs on falling edge of signal
1 - External interrupt occurs on rising edge of signal

(P12C50x) **Enable weak pull-ups**
0 - Enabled: (GP0, 1, and 3 float height)
1 - Disabled

(P16C5x) **Unassigned**

(P16C71) **Port B pull-up enable**
0 - Disabled
1 - Enabled

(P12C50x) **Enable wake-up on pin change**
0 - Enabled: (A change in GP0, 1, or 3 wakes the PIC up from sleep)
1 - Disabled

STATUS

Bit no.	7	6	5	4	3	2	1	0
Bit name	PA2	PA1	PA0	TO	PD	Z	DC	C

Carry/borrow flag:
Reacts to carrying or borrowing with arithmetic operations, and to the **rrf** and **rlf** instructions.

Digit carry/borrow flag: As carry flag except concerning the lower nibbles of numbers in question.

Zero flag:
1: The result was 0
0: The result wasn't 0

Power Down and **TimeOut** bits. See Table C.1

(P16C54 and 55)
Do not use these bits for anything, in order to maintain upward compatibility.

(P16C56 and 57)
00 - Page 0 (000–1FF)
01 - Page 0 (200–3FF)
10 - Page 0 (400–5FF)
11 - Page 0 (600–7FF)

(P12C7x)
00 - Bank 0
01 - Bank 1

Do not use these bits for anything, in order to maintain upward compatibility.

Table C.1 Power Down and TimeOut bits

TO	PD	Reset caused by...
0	0	WDT wakeup from sleep
0	1	WDT timeout (not during sleep)
1	0	MCLR wakeup from sleep
1	1	Power-up

INTCON

Bit no.	7	6	5	4	3	2	1	0
Bit name	GIE	ADIE	T0IE	INTE	RBIE	T0IF	INTF	RBIF

RB Change flag:
1: One of the bits RB4–RB7 have changed.
0: None of these bits have changed.
(**Note:** Must be cleared by you)

Ext. int. flag:
1: The ext. interrupt has occurred.
0: The ext. interrupt has not occurred.

TMR0 interrupt flag:
1: TMR0 has overflowed.
0: TMR0 hasn't overflowed.
(**Note:** Must be cleared by you)

RB change enable:
1: Enables RB change interrupt.
0: Disables it.

Ext. int. enable:
1: Enables ext. interrupt.
0: Disables it.

TMR0 interrupt enable:
1: Enables TMR0 interrupt.
0: Disables it.

A/D conversion interrupt enable:
1: Enables A/D interrupt.
0: Disables it.

Global interrupt enable:
1: Enables all interrupt which have been selected above.
0: Disables all interrupts.

ADCON0

Bit no.	7	6	5	4	3	2	1	0
Bit name	ADCS1	ADCS0	-	CHS1	CHS0	GO	ADIF	ADON

A/D on bit:
1: A/D converter is on.
0: A/D converter is off (and consumes no operating current).

A/D int. flag:
1: The A/D interrupt has occurred.
0: The A/D interrupt has not occurred. Must be cleared by you.

GO/DONE:
1: Conversion is in progress. Setting this bit starts conversion.
0: Conversion has finished.

Channel select: Selects which input to read.
00: RA0/AN0
01: RA1/AN1
10: RA2/AN2
11: RA3/AN3

Reserved: Do what you want with it!

A/D clock select: Selects how long the PIC takes over a A/D conversion
00: Fosc/2
01: Fosc/8
10: Fosc/32
11: FRC

ADCON1

Bit 1	Bit 0	RA0	RA1	RA2	RA3	Ref
0	0	A	A	A	A	VDD
0	1	A	A	A	Vref	RA3
1	0	A	A	D	D	VDD
1	1	D	D	D	D	VDD

Appendix D
If all else fails, read this

You should find that there are certain mistakes which you make time and time again (I do!). I've listed the popular ones here:

1. Look for: **subwf**   **FileReg**   ... are you sure you don't mean ...

   subwf   **FileReg, w**

2. *You are using the PIC71* ... have you remembered that the general purpose file registers start from address **0C** and *not* from **08** as in the PIC5x series?

3. *You are using the PIC71* ... have you remembered to set up bit 7 of the OPTION register correctly?

4. Are your subroutines in the correct page or half of page?

5. Are you adding something to the program counter in the wrong place on a page or on the wrong page?

6. Are you remembering to reset a file register you are using to keep track of how many times something has happened (e.g. a postscaler)?

7. <u>If you are having a total nightmare and NOTHING is working</u>... have you specified the correct PIC at the top of the program?

Appendix E
Some useful contact information

John Morton: JJLMorton@aol.com

Microchip: +1 (602) 7886 7200
(Corporate Office)   Microchip Technology Inc.
(The guys that make PICs)   2355 West Chandler Blvd.
                                   Chandler
                                   AZ 85224-6199
                                   USA

Bluebird Electronics   (01380) 725 110
(Run by Nigel Gardner – a PIC
expert who runs training courses)

Microchip   (01509) 611 344
(UK Helpline)
(Really friendly assistance)

LearnPIC website   http://members.aol.com/LearnPIC
(Regularly updated PIC website)

Appendix F
Ordering project PCBs

Boards may be ordered singly at the following prices:

Projects	Price
A, B, C, D	£7.00
E	£7.00
F, G, H	£7.00
I	£11.00
J	£7.00
K	£6.50
L	£9.50
M, N	£6.50
O	£11.00
P	£5.00
Complete set	**£65.00**

Order form

Projects	Quantity	Subtotal
A, B, C, D	☐	☐
E	☐	☐
F, G, H	☐	☐
I	☐	☐
J	☐	☐
K	☐	☐
L	☐	☐
M, N	☐	☐
O	☐	☐
P	☐	☐
Complete set	☐	☐

TOTAL: _____

Send cheque payable to Radley College:

Electronics Department
Radley College
Abingdon
Oxon
OX14 2HR

Appendix G
References and
further reading

1. Gardner, N. and Birnie, P. (1997). An Introduction to the ICEPIC, In *PIC Cookbook, Volume 2,* pp.145–154, Bluebird Technical Press Limited. **(This chapter explains the basics on how to use the ICEPIC emulator.)**

2. Gardner, N. and Birnie, P. (1995). *PIC Cookbook*

3. Gardner, N. and Birnie, P. (1997). *PIC Cookbook, Volume 2.* Bluebird Technical Press Limited. **(A series which is packed with example PIC programs and program segments.)**

4. Gardner, N. (1996). *A Beginner's Guide to the Microchip PIC.* Bluebird Technical Press Limited.

5. Microchip (1996). *PIC16/17 Microcontroller Data Book.* **(Raw information (and lots of it) on all the PICs available.)**

6. Gardner, N. and Horsey, M. (1996). *From Bits to Chips.* Bluebird Electronics.

7. *Everyday and Practical Electronics.* **(A monthly magazine which normally has a PIC project or two.)**

Appendix H
Answers to the exercises

Chapter 1 Introduction

1.1(a)   Largest power of two less than 234 = 128 = 27. Bit 7 = 1
          This leaves 234 – 128 = 106.   64 is less than 106 so bit 6 = 1,
          This leaves 106 – 64 = 42.   32 is less than 42 so bit 5 = 1,
          This leaves 42 – 32 = 10.   16 is greater than 10 so bit 4 = 0,
                                       8 is less than 10 so bit 3 = 1
          This leaves 10 – 8 = 2   4 is greater than 2 so bit 2 = 0,
                                       2 equals 2 so bit 1 = 1
          Nothing left so bit 0 = 0.

          The resulting binary number is **11101010.**

   (b) OR ...
          Divide 234 by two.   Leaves 117, remainder **0**
          Divide 117 by two.   Leaves 58, remainder **1**
          Divide 58 by two.   Leaves 29, remainder **0**
          Divide 29 by two.   Leaves 14, remainder **1**
          Divide 14 by two.   Leaves 7, remainder **0**
          Divide 7 by two.   Leaves 3, remainder **1**
          Divide 3 by two.   Leaves 1, remainder **1**
          Divide 1 by two.   Leaves 0, remainder **1**

          So **11101010** is the binary equivalent.

1.2 (a)   Largest power of two less than 157 = 128 = 27. Bit 7 = 1,
          This leaves 157 – 128 = 29.   64 is greater than 29 so bit 6 = 0,
                                       32 is greater than 29 so bit 5 = 0,
                                       16 is less than 29 so bit 4 = 1,
          This leaves 29 – 16 = 13   8 is less than 13 so bit 3 = 1,
          This leaves 13 – 8 = 5   4 is less than 5 so bit 2 = 1,
          This leaves 5 – 4 = 1   2 is greater than 1 so bit 1 = 0,
                                       1 equals 1 so bit 0 = 1.

          The resulting binary number is **10011101.**

(b) OR...

Divide 157 by two.	Leaves 78, remainder **1**
Divide 78 by two.	Leaves 39, remainder **0**
Divide 39 by two.	Leaves 19, remainder **1**
Divide 19 by two.	Leaves 9, remainder **1**
Divide 9 by two.	Leaves 4, remainder **1**
Divide 4 by two.	Leaves 2, remainder **0**
Divide 2 by two.	Leaves 1, remainder **0**
Divide 1 by two.	Leaves 0, remainder **1**

So **10011101** is the binary equivalent.

1.3 There are 14 16s in 234, leaving 234 – 224 = 10. So bit 1 = 14 = D, and bit 0 = 10 = A. The number is therefore **DA**.

1.4 There are 9 16s in 157, leaving 157 – 144 = 13. So bit 1 = 9, and bit 0 = 13 = D. The number is therefore **9D**.

1.5 1110 = 14 = E. 1010 = 10 = A. The number is therefore **EA**.

1.6 1. One push button requires **one** input.
2. Four seven-segment displays require 4 + 7 = **11** outputs, creating a total of **12 I/O pins** which will just fit onto a **PIC54.**

1.7

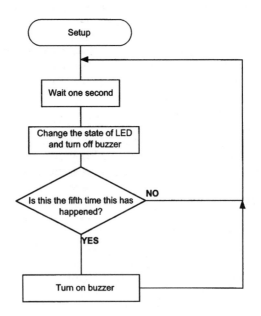

1.8   b'0001'
        b'0010'
        b'0100'
        b'1000'
        b'0001'   ...and so on.

1.9
Init   clrf   porta
        clrf   portb
        clrf   portc
        movlw   b'1001'
        tris   porta
        movlw   b'10000000'
        tris   portb
        movlw   b'01111100'
        tris   portc
        retlw   0

Chapter 2 Exploring the PIC5x series

2.1 Bits 6 and 7 are always 0.
The TMR0 is counting *externally*, so bit 5 is 1.
It's irrelevant whether the TMR0 is *rising* or *falling edge triggered* so bit 4 is 0 or 1.
No prescaling for the TMR0 is required, so bit 3 is 1.
WDT is not be used, so WDT prescaling is irrelevant.

The number to be moved into the option register is **00100000**.

2.2 **Challenge!**
Bits 6 and 7 are always 0.
The TMR0 is counting *externally*, so bit 5 is 1.
It's irrelevant whether the TMR0 is *rising* or *falling edge triggered* so bit 4 is 0 or 1, (let's say 0).
Prescaling for the TMR0 is required, so bit 3 is 0.
$256 \times 4 = 1024$, so prescaling of 4 is required, so bits 2 is 0, bit 1 is 0, and bit 0 is 1.

The number to be moved into the option register is **00100001.**

2.3
        movlw   b'10101000'   ; moves the correct number into the
                                    ;   working reg.
        xorwf   portb   ; toggles the correct bits in portb

2.4 The **,f** or **,w** after the specified file register (e.g. **comf   porta, f**) selects the destination of the instruction result. **,f** leaves the result in the file register and **,w** puts the result in the working register, leaving the file register unchanged.

2.5

```
        movlw   b'00010100'   ; motorists: green on, others off
        movwf   portb         ; pedestrians: red on, others off
```

2.6

```
ButtonLoop  btfss   porta, 0       ; is the pedestrians' button pressed?
            goto    ButtonLoop     ; no, so loops back
```

2.7

```
        bsf     portb, 1      ; turns motorists' amber light on
        bcf     portb, 2      ; turns motorists' green light off
```

2.8

```
        call    delay         ; waits half a second
        call    delay         ; waits half a second
        call    delay         ; waits half a second
        call    delay         ; waits half a second
```

2.9

```
        movlw   b'00100001'   ; motorists: red on, amber off
        movwf   portb         ; pedestrians: green on, red off
```

2.10

```
        bsf     portb, 1      ; turns motorists' amber light on
        bcf     portb, 0      ; turns motorists' red light off
```

2.11 **Challenge!**

```
            movlw   d'8'           ; moves the decimal number 8 into
            movwf   Counter8       ;   Counter8
FlashLoop   movlw   b'00100000'    ; toggles the green pedestrian light
            xorwf   portb, f       ;
            call    delay          ; creates half second delay
            decfsz  Counter, f     ; makes it happen eight times
            goto    FlashLoop      ; loops back
```

2.12

```
delay   movlw   d'15'         ; moves the decimal
        movwf   Mark15        ;   number 15 into the GPF called
                              ;   Mark15
        movlw   d'80'         ; moves the decimal
```

	movwf	Post80	; number 80 into the GPF
			; called Post80
TimeLoop	movfw	Mark15	; takes the number out of
			; Mark15
	subwf	TMR0, w	; subtracts this number from
			; the number inTMR0,
			; leaving the result in the
			; working
			; register (and leaving TMR0
			; unchanged)
	btfss	STATUS, Z	; tests the zero flag - skip if set,
			; i.e. if the result is zero it
			; will skip the next instruction
	goto	TimeLoop	; if the result isn't zero, it
			; loops back to 'Loop'
	movlw	d'15'	; moves the decimal number
	addwf	Mark15, f	; 15 into the working register
			; and then adds it to 'Mark15'
	decfsz	Post80, f	; decrements 'Post80', and
			; skips the next instruction if
			; the result is zero
	goto	TimeLoop	; if the result isn't zero, it
			; loops back to 'Loop'
	retlw	0	; when half a second has
			; passed, it returns

2.13 dacgbfe

0	**11101110**
1	**00101000** or **00000110**
2	**11011010**
3	**11111000**
4	**00111100**
5	**11110100**
6	**11110110**
7	**01101000**
8	**11111110**
9	**11111100**
A	**01111110**
b	**10110110**
c	**10010010**

d	**10111010**
E	**11010110**
F	**01010110**

2.14 In both cases the number being moved into the program counter is **0043**.

2.15

Main	**btfss**	**portb, 0**	; tests push button
	goto	**Main**	; if not pressed, loops back

2.16

	incf	**Counter, f**	;

2.17 Challenge!

	movlw	**d'16'**	; has Counter reached 16?
	subwf	**Counter, w**	;
	btfsc	**STATUS, Z**	;
	clrf	**Counter**	; if yes, resets Counter

2.18

	movfw	**Counter**	; moves Counter into the
			; working reg.
	call	**_7SegDisp**	; converts into 7 seg. code
	movwf	**portb**	; displays value
	goto	**Main**	; loops back to Main

2.19
_7SegDisp

	addwf	**PCL**	; skips a certain number of
			; instructions
	retlw	**b'11111110'**	; code for 0
	retlw	**b'01100000'**	; code for 1
	retlw	**b'11011010'**	; code for 2
	retlw	**b'11110010'**	; code for 3
	retlw	**b'01100110'**	; code for 4
	retlw	**b'10110110'**	; code for 5
	retlw	**b'10111110**	; code for 6
	retlw	**b'11100000'**	; code for 7
	retlw	**b'11111110'**	; code for 8
	retlw	**b'11110110'**	; code for 9
	retlw	**b'11101110'**	; code for A
	retlw	**b'00111110'**	; code for b
	retlw	**b'10011100'**	; code for C
	retlw	**b'01111010'**	; code for d

```
          retlw      b'10011110'      ; code for E
          retlw      b'10001110'      ; code for F
```

2.20

```
TestLoop  btfss      portb, 0         ; tests push button
          b          Main             ; released, so returns
          b          TestLoop         ; still pressed, so keeps looping
```

2.21

```
          clrf       TMR0             ; resets TMR0
TimeLoop  movlw      d'255'           ; has TMR0 reached 255?
          subwf      TMR0, w          ;
          btfss      STATUS, Z        ;
          goto       TimeLoop         ; if not, keeps looping
          goto       Main             ; 0.07 seconds has passed, so
                                      ;   goes to Main
```

2.22

```
          movlw      d'10'            ; tests to see whether Seconds
          subwf      Seconds, w       ;   has reached 10 (i.e. whether
                                      ;   or not ten seconds have
          btfss      STATUS, Z        ;   passed)
          retlw      0                ; 10 seconds haven't passed, so
                                      ;   returns

          clrf       Seconds          ; 10 seconds have passed, so
          incf       TenSecond, f     ;   resets Seconds and
                                      ;   increments the number of
                                      ;   tens of seconds

          movlw      d'6'             ; tests to see whether
          subwf      TenSecond, w     ;   TenSecond has reached 6
                                      ;   (i.e. whether or not one
          btfss      STATUS, Z        ;   minute has passed)
          retlw      0                ; 1 minute hasn't passed, so
                                      ;   returns

          clrf       TenSecond        ; 1 minute has passed, so resets
          incf       Minutes, f       ;   TenSecond and increments
                                      ;   the number of minutes
```

2.23 The number required would be **00000011**.

2.24

Display1	**movfw**	**Seconds**	**; takes the number out of**
			; Seconds
	call	**_7SegDisp**	**; converts the number into 7**
			; seg. code
	movwf	**portb**	**; displays the value through**
			; Port B
	movlw	**b'0001'**	**; turns on correct display**
	movwf	**porta**	**;**
	retlw	**0**	**; returns**
Display10	**movfw**	**TenSecond**	**; takes the number out of**
			; TenSecond
	call	**_7SegDisp**	**; converts the number into 7**
			; seg. code
	movwf	**portb**	**; displays the value through**
			; Port B
	movlw	**b'1000'**	**; turns on correct display**
	movwf	**porta**	**;**
	retlw	**0**	**; returns**
DisplayMin	**movfw**	**Minutes**	**; takes the number out of**
			; Minutes
	call	**_7SegDisp**	**; converts the number into 7**
			; seg. code
	movwf	**portb**	**; displays the value through**
			; Port B
	movlw	**b'0100'**	**; turns on correct display**
	movwf	**porta**	**;**
	retlw	**0**	**; returns**

2.25

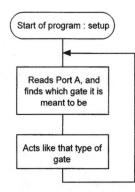

2.26 The resulting number would be **00010000**.

2.27

Main	btfsc	**porta, 1**	; tests setting switch
	goto	**GreenLed**	; switch is high, so turn on
			; green LED
	bsf	**portb, 1**	; switch is low, so turn on red
			; LED

2.28

GreenLed	bsf	**portb, 0**	; turns on green LED
	goto	**TenthSecond**	; loops back to main body of
			; program.

2.29

TenthSecond	movfw	**TMR0**	; is TMR0 at 0?
	btfss	**STATUS, Z**	;
	goto	**Continue**	;
	incf	**TMR0**	;
	decfsz	**Post256**	;
	goto	**Continue;**	
	clrf	**portb**	; it has, so turns off all LEDs
Continue	etc...		

2.30

	btfsc	**porta, 1**	; tests setting switch
	goto	**TenthSecond**	; disabled, so doesn't test
			; trigger input

2.31

	btfss	porta, 0	; tests to see whether motion
			; sensor has been set
	got	TenthSecond	; not triggered, so loops back

2.32

	bsf	porta, 3	; turns on siren
EndLoop	clrwdt		; resets watchdog timer
	goto	EndLoop	; constantly loops

2.33

Start	btfsc	STATUS, 4	; we need only test the
			; TimeOut bit
	call	PreInit	; set, so calls subroutine
	etc.		; clear, so skips subroutine

Chapter 3 Exploring the PIC71

3.1

	movlw	b'10001000'	; enables RBChange and
	movwf	INTCON	; global interrupts
	sleep		; goes to sleep
	nop		; the line after sleep is always
			; executed, and we want
			; nothing to happen
	bsf	INTCON, 4	; enables also the external
			; interrupt

3.2

Init	clrf	porta	; resets input/output ports
	clrf	portb	
	goto	InitContinue	; skips address 0004
	goto	isr	; at address 0004, goes to isr
InitContinue			
	bsf	STATUS, 5	; selects bank 1
	clrf	TRISA	; RA0: buzzer, RA1–RA4: not
			; connected
	movlw	b'11110000'	; RB0–RB3: LEDs, RB4–RB7:
	movwf	TRISB	; push buttons
	movlw	b'00000111'	; sets up timing register
	movwf	OPTION	;
	bcf	STATUS, 5	; goes back to bank 0
	movlw	b'10001000'	; enables RBChange, and
	movwf	INTCON	; global interrupts

```
                return                              ; returns from interrupt

3.3
                swapf      portb, f                 ; turns appropriate LED on
                bsf        porta, 0                 ; turns on buzzer

3.4
                movlw      b'00100000'              ; enables TMR0 interrupt,
                movwf      INTCON                   ;    disables RBChange
                retfie                              ; returns, setting the global
                                                    ;    interrupt enable

3.5
Init            clrf       porta                    ; resets input/output ports
                clrf       portb
                goto       InitContinue             ; skips address 0004
                goto       isr                      ; at address 0004, goes to isr
InitContinue
                bsf        STATUS, 5                ; selects bank 1
                movlw      b'00001'                 ; RA0: temperature sensor,
                movwf      TRISA                    ;    RA1–4: not connected
                clrf       TRISB                    ;    RB0–2: LEDs, RB3–7: not
                                                    ;        connected

                movlw      b'00000010'              ; sets up ADCON1: RA0 and
                movwf      ADCON1                   ;    RA1 as analogue inputs,
                                                    ;    Vref is the supply voltage

                movlw      b'00000111'              ; sets up timing register
                movwf      OPTION                   ;
                bcf        STATUS, 5                ; goes back to bank 0

                movlw      b'01000001'              ; sets up A/D register:
                movwf      ADCON0                   ; clock: Fosc/8, channel: AN0,
                                                    ;    converter is on

                movlw      b'11000000'              ; sets up interrupts register:
                movwf      INTCON                   ; global and A/D
                                                    ;    interrupts enabled

                return                              ; returns from the
                                                    ;    subroutine
```

3.6

	bcf	ADCON0, 1	; resets A/D interrupt flag

3.7

Cold	movlw	b'00000001'	; turns on cold LED
	movwf	portb	; (others off)
	retfie		; returns, enabling global
			; interrupt

Ok	movlw	b'00000010'	; turns o.k. LED (others off)
	movwf	portb	;
	retfie		; returns, enabling global
			; interrupt

Hot	movlw	b'00000100'	; turns on hot LED
	movwf	portb	; (others off)
	retfie		; returns, enabling global
			; interrupt

3.8

Init	clrf	porta	; resets I/O ports
	clrf	portb	
	b	Init+4	; leaves address 0004 for isr
	b	isr	
	bsf	STATUS,5	; selects bank 1
	movlw	b'00001'	; RA0: contacts, RA1– RA3: 7
	movwf	TRISA	; seg. controlling pins
	movlw	b'00000001'	; RB0: mode button,
	movwf	TRISB	; RB1–RB7: 7 seg. code
	movlw	b'00000111'	; sets up TMR0, prescaled at
	movwf	OPTION	; 256
	movlw	b'00000010'	; RA0 and RA1 are analogue,
	movwf	ADCON1	; Vref is supply voltage
	bcf	STATUS,5	; returns to bank 0
	movlw	b'00010000'	; sets up interrupts:
	movwf	INTCON	; external on, global off
	movlw	b'01000001'	; sets up A/D conversion:
	movwf	ADCON0	; clock: Fosc/8, channel: AN0,
			; converter is on

	return		; returns from subroutine

3.9

	incf	Chooser	; goes on to next mode
	movlw	d'5'	; has it gone through all five
	subwf	Chooser,w	; modes?
	btfsc	STATUS,Z	;
	clrf	Chooser	; yes, so reset back to mode 1
			; (1-6)

3.10

	movfw	Chooser	; takes the number out of
			; Chooser
	addwf	PCL, f	; skips that many
			; instructions
	b	Dis6	; 1-6_
	b	DisLot	; Lott
	b	Dis12	; 1-12
	b	Dis99	; 1-99
	b	Dis66	; -6-6

3.11

Dis6	movlw	b'10111110'	; 6
	movwf	dig3	
	movlw	b'00000000'	; blank
	movwf	dig4	
	b	DisCommon	; goes to common place to set
			; up 1 -

Dis12	movlw	b'11011010'	; 2
	movwf	dig4	
	movlw	b'01100000'	; 1
	movwf	dig3	
	b	DisCommon	;

Dis99	movlw	b'11100110'	; 9
	movwf	dig3	
	movlw	b'11100110'	; 9
	movwf	dig4	
	b	DisCommon	;

3.12

DisCommon

	movlw	b'01100000'	; 1
	movwf	dig1	

```
                movlw      b'00000010'     ; -
                movwf      dig2
                return                      ; returns, without re-enabling
                                            ;   the interrupts
3.13
DisLot          movlw      b'00011100'     ; L
                movwf      dig1
                movlw      b'00111010'     ; o
                movwf      dig2
                movlw      b'00001110'     ; t
                movwf      dig3
                movlw      b'00001110'     ; t
                movwf      dig4
                return


Dis66           movlw      b'00000010'     ; sets a dash
                movwf      dig1
                movwf      dig3
                movlw      b'10111110'     ; sets a number 6
                movwf      dig2
                movwf      dig4
                return


3.14
Start
                call       Init            ; set everything up
                bsf        INTCON, 1       ; tell the isr to go to mode
                                           ;   changing section
                call       isr             ; call isr


3.15
Main            bsf        ADCON0,2        ; starts A/D conversion
                call       display         ; sorts out displays
                btfsc      portb, 0        ; tests for up button
                b          ADTest          ; still pressed, so changes
                                           ;   nothing
                movlw      b'10100000'     ; released, so enables TMR0
                                           ;   and global
                movwf      INTCON          ;   and global interrupt


3.16
ADTest          btfsc      ADCON0, 2       ; conversion is in progress
                b          Main+1          ; goes back to main, and skips
                                           ;   one line
```

```
          movlw    d'10'          ; is result less than 10? (are
          subwf    ADRES, w       ;    contacts pressed?)
          btfss    STATUS, C      ;
          b        Main           ; no, so loops back
          etc.                    ; yes, so continues
```

3.17

```
          btfss    INTCON, 1      ; has external interrupt
                                  ;    occurred?
          b        TMRInt         ; no, so goes to timer interrupt
          etc.                    ; yes, so continues
```

3.18

```
TMRInt    bsf      INTCON, 4      ; enables external interrupt

          decfsz   Post256        ; first postscaled by 256
          retfie                  ;
          decfsz   Post7          ; secondly postscaled by 7, (3.5
                                  ;    minutes passed?)
          retfie                  ; no

          movlw    d'7'           ; yes, so resets second
          movwf    Post7          ;    postscaler
          clrf     portb          ; clears outputs
          clrf     porta          ;
          sleep                   ; sleeps (low power mode)
          retfie                  ; returns, enabling the global
                                  ;    interrupt
```

3.19

```
          clrf     Post256        ; resets sleeping postscalers
          movlw    d'7'
          movwf    Post7
          bcf      INTCON, 7      ; no more interrupts from now
                                  ;    on
```

3.20

Stabilizer

```
          bsf      ADCON0, 2      ; starts conversion
          btfsc    ADCON0, 2      ; has it finished?
          b        Stabilizer     ; no, so keeps looping

          decfsz   ADCount        ; has this happened 256 times
          b        Stabilizer     ; no, so loops back
```

```
              movfw      ADRES            ; yes, so stores A/D result
              movwf      Skin

3.21
Main2         bsf        ADCON0,2         ; starts A/D conversion (are
                                          ;    contacts released?)
              call       Chaser           ; sorts out chasing while
                                          ;    contacts are pressed
              call       display          ;

              btfsc      ADCON0,2         ; is conversion finished?
              b          Main2+1          ; no, so loops back

3.22
              movlw      d'10'            ; is A/D low enough to suggest
              subwf      ADRES,w          ;    finger has been removed?
              btfsc      STATUS,C         ;
              b          Main2            ; no, so loops back an starts
                                          ;    another conversion

              decfsz     LetGoCount       ; yes, confirms low result 200
                                          ;    times
              b          Main2            ;
              movlw      d'200'           ; resets the counting GPF
              movwf      LetGoCount       ;    register

3.23
              movfw      TMR0             ; takes the number out of
                                          ;    TMR0
              movwf      TouchTime        ; stores it in TouchTime

3.24
              call       MessageChooser   ; choose message using
                                          ;    number in Skin
              movlw      d'28'            ; resets first postscaler, so that
              movwf      Post28           ;    the PIC times the full
                                          ;    3 seconds
              bsf        sec3             ;

Loop3Sec      movlw      b'10100000'      ; enables TMR interrupt
              movwf      INTCON           ;
              call       display          ; keeps displays going
              btfsc      sec3             ;
              b          Loop3Sec         ; keeps looping until 3
                                          ;    seconds have passed
```

3.25

movlw	d'28'	; resets postscaler
movwf	Post28	;
bsf	sec3	;

3.26

movfw	TouchTime	; adds TouchTime and Skin
		; together, storing the
addwf	Skin, w	; result in the GPF, Random
movwf	Random	;
movfw	Chooser	; use the mode to jump to the
addwf	PCL	; correct section
b	Ran6	
b	RanLot	
b	Ran12	
b	Ran99	
b	Ran66	

3.27

Ran6	movlw	d'6'
	b	Adder
Ran12	movlw	d'12'
	b	Adder
Ran99	movlw	d'99'
	b	Adder

3.28

Adder	movwf	Scaler	; stores the number from
			; Ran6, Ran12, or Ran99
	movfw	Scaler	; repeatedly adds the number
	addwf	Random	; until it passes 255
	btfss	STATUS,C	; is the value suitable?
	b	Adder+1	; no, so loops back
	incf	Random	; adds one to Random

3.29

	movwf	Tens	; stores as first 1-6 value
	addwf	TouchTime	; gets new random number
	movwf	Random	;
Ran66	movlw	d'6'	; converts it into 1-6
	addwf	Random	;
	btfss	STATUS, C	;
	b	Ran66	;

```
                incf        Random          ;
                b           Continue66      ; jumps to correct place in
                                            ;   TensLoop
```

3.30

```
RanLot          movlw       d'49'           ; repeatedly adds 49 until it
                addwf       Random          ;   goes past 255
                btfss       STATUS,C        ; is number suitable?
                b           RanLot          ; no, so loops back
                incf        Random          ; adds on, so it is between 1
                                            ;   and 49
```

3.31 Challenge!
CompareLott

```
                movlw       1Bh             ; selects Lott1 first
                movwf       FSR
                movfw       INDF            ; compares with previous
                subwf       Random, w       ;   lottery values
                btfsc       STATUS,Z        ;
                b           Changer         ; the two are the same, so gets
                                            ;   new number
```

3.32

```
Changer         movfw       Skin            ; gets new random number
                addwf       Random          ;
                b           RanLot          ; converts it to a number
                                            ;   between 1 and 49
```

3.33

```
                incf        FSR             ; moves on to next lottery
                                            ;   number
                btfss       FSR, 5          ; has it gone too far?
                b           CompareLott     ; no, so loops back
                etc.                        ; yes, so continues
```

3.34

```
                movfw       LottCount       ; use number from LottCount,
                addlw       1B              ;   and add 1B to it in order to
                                            ;   select the correct GPF
                movwf       FSR             ;
                movfw       Random          ;
                movwf       INDF            ;
                incf        LottCount       ; moves on to next lottery
                                            ;   number
```

3.35

TensLoop	incf	Tens	; works the tens digit of the
			; number
	movlw	d'10'	; (keeps subtracting 10 until
			; result is negative)
	subwf	Random	;
	btfsc	STATUS,C	;
	b	TensLoop	;

3.36

	movlw	d'10'	; gets units value by adding 10
	addwf	Random, w	; to Random and leaving the
			; result in the working
			; register.
Continue66	call	Decoder	; converts units value into 7
			; seg. code
	movwf	dig4	; moves code into appropriate
			; digit
	clrf	dig3	; blanks out left three digits
	clrf	dig2	;
	clrf	dig1	;
Decoder	addwf	PCL	; converts number into 7 seg.
			; code
	retlw	b'11111100'	; number 0
	retlw	b'01100000'	; number 1
	retlw	b'11011010'	; number 2
	retlw	b'11110010'	; etc.
	retlw	b'01100110'	;
	retlw	b'10110110'	;
	retlw	b'10111110'	
	retlw	b'11100000'	;
	retlw	b'11111110'	;
	retlw	b'11100110'	;

3.37

	movfw	Tens	
	btfsc	STATUS, Z	; checks to see if tens is zero
	b	LottCounter	; yes, so displays nothing on
			; dig3
	call	Decoder	; converts result into 7 seg.
			; code
	btfss	Chooser, 2	; is the device in -6-6 mode?

	b	Not66	; goes to a place where the tens ; is shown on dig3
	movwf	dig1	; in -6-6 mode, so moves code ; into dig1
	b	LottCounter	; now goes to label the lottery ; number correctly

Not66	movwf	dig3	; moves code into dig 3
LottCounter	etc.		

3.38

LottCounter	clrf	Tens	; resets Tens register
	decfsz	Chooser, w	; is the device in lottery mode?
	b	Main	; no, so loops back to main
	etc.		; yes, so continues

3.39

	movlw	b'00000010'	; displays a dash on digit 2
	movwf	dig2	;
	movfw	LottCount	; takes the number out of
	call	Decoder	; LottCount and uses it ; in dig1
	movwf	dig1	;

3.40

display	movfw	TMR0	; use TMR0 to select one of
	andlw	b'00000011'	; four sections
	addwf	PCL	;
	b	digit1	
	b	digit2	
	b	digit3	
	b	digit4	

digit1	movlw	b'10000'	; turns on relevant display
	movwf	porta	;
	movfw	dig1	
	movwf	portb	; moves relevant code into port ; b
	return		

digit2	movlw	b'00010'	; turns on relevant display
	movwf	porta	;
	movfw	dig2	
	movwf	portb	; moves relevant code into port

			; b
	return		
digit3	movlw	b'00100'	; turns on relevant display
	movwf	porta	;
	movfw	dig3	
	movwf	portb	; moves relevant code into port
			; b
	return		
digit4	movlw	b'01000'	; turns on relevant display
	movwf	porta	;
	movfw	dig4	
	movwf	portb	; moves relevant code into port
			; b
	return		

3.41

Chaser passed?	movfw	Mark240	; has a tenth of a second
	subwf	TMR0, w	;
	btfss	STATUS, Z	;
	return		; no, so returns
	movlw	d'240'	; yes, so continues and resets
	addwf	Mark240	; marker

3.42

	First	Second	Third
dig1 :	00000100	10010000	00001000
dig2 :	00010000	00000000	10000000
dig3 :	10000000	00000000	00010000
dig4 :	00100000	10010000	01000000

3.43

	incf	ChaseCount	; scrolls through the 3
			; different displays
	movfw	ChaseCount	; codes, creating a chase
	addwf	PCL	
	b	Chase1	
	b	Chase2	
	b	Chase3	
Chase1	movlw	b'00000100'	; sets up first chase pattern
	movwf	dig1	;

```
                movlw      b'00010000'
                movwf      dig2
                movlw      b'10000000'
                movwf      dig3
                movlw      b'00100000'
                movwf      dig4
                return

Chase2          movlw      b'10010000'      ; sets up second chase
                movwf      dig1             ;   pattern
                movlw      b'00000000'
                movwf      dig2
                movlw      b'00000000'
                movwf      dig3
                movlw      b'10010000'
                movwf      dig4
                return

Chase3          movlw      b'00001000'      ; sets up third chase
                movwf      dig1             ;   pattern
                movlw      b'10000000'
                movwf      dig2
                movlw      b'00010000'
                movwf      dig3
                movlw      b'01000000'
                movwf      dig4

                movlw      d'255'           ; resets ChaseCount GPF
                movwf      ChaseCount       ;   register
                return
```

3.44

MessageChooser

```
                movlw      d'12'            ; is Skin between 11 and 12?
                subwf      Skin,w           ;
                btfss      STATUS, C        ;
                b          sad              ; yes, so displays SAd
```

3.45

```
                movlw      d'15'            ; is Skin between 13 and 15?
                subwf      Skin,w           ;
                btfss      STATUS, C        ;
                b          bad              ; yes, so displays bAd

                movlw      d'20'            ; is Skin between 16 and 20?
```

```
             subwf     Skin,w       ;
             btfss     STATUS, C    ;
             b         cool         ; yes, so displays cool

             movlw     d'25'        ; is Skin between 21 and 25?
             subwf     Skin,w       ;
             btfss     STATUS, C    ;
             b         john         ; yes, so displays John
             movlw     d'35'        ; is Skin between 26 and 35?
             subwf     Skin,w       ;
             btfss     STATUS, C    ;
             b         hot          ; yes, so displays hot

             movlw     d'50'        ; is Skin between 36 and 50?
             subwf     Skin,w       ;
             btfss     STATUS, C    ;
             b         tops         ; yes, so displays to PS
             b         ace          ; no, so displays ACE (above
                                    ;    50)
```

3.46

```
_A           retlw     b'11101110'  ; letter A
_b           retlw     b'00111110'  ; letter b
_C           retlw     b'10011100'  ; letter C
_c           retlw     b'00011010'  ; letter c
_d           retlw     b'01111010'  ; letter d
_E           retlw     b'10011110'  ; E
_F           retlw     b'10001110'  ; F
_g           retlw     b'11110110'  ; g
_H           retlw     b'01101110'  ; H
_h           retlw     b'00101110'  ; h
_I           retlw     b'00001100'  ; I
_i           retlw     b'00001000'  ; i
_j           retlw     b'01110000'  ; j
_L           retlw     b'00011100'  ; L
_n           retlw     b'00101010'  ; n
_O           retlw     b'11111100'  ; O
_o           retlw     b'00111010'  ; o
_P           retlw     b'11001110'  ; P
_q           retlw     b'11100110'  ; q
_r           retlw     b'00001010'  ; r
_S           retlw     b'10110110'  ; S
_t           retlw     b'00001110'  ; t
_U           retlw     b'01111100'  ; U
```

```
_u              retlw           b'00111000'        ; u
_y              retlw           b'01110110'        ; y
blank           retlw           b'00000000'        ; blank


3.47
sad             call            _S
                movwf           dig1
                call            _A
                movwf           dig2
                call            _d
                movwf           dig3
                call            blank
                movwf           dig4
                return


bad             call            _b
                movwf           dig1
                call            _A
                movwf           dig2
                call            _d
                movwf           dig3
                call            blank
                movwf           dig4
                return


cool            call            _C
                movwf           dig1
                call            _O
                movwf           dig2
                call            _O
                movwf           dig3
                call            _L
                movwf           dig4
                return


john            call            _j
                movwf           dig1
                call            _o
                movwf           dig2
                call            _h
                movwf           dig3
                call            _n
                movwf           dig4
                return
```

```
hot        call      _h
           movwf     dig1
           call      _o
           movwf     dig2
           call      _t
           movwf     dig3
           call      blank
           movwf     dig4
           return

tops       call      _t
           movwf     dig1
           call      _o
           movwf     dig2
           call      _P
           movwf     dig3
           call      _S
           movwf     dig4
           return

ace        call      _A
           movwf     dig1
           call      _C
           movwf     dig2
           call      _E
           movwf     dig3
           call      blank
           movwf     dig4
           return
```

Index